You Can Do Chemistry:
Moles & Stoichiometry

Copyright & Other Notices

Published in 2018 by Answers 2000 Limited

Copyright © 2018, Sunil Tanna

Sunil Tanna has asserted his right to be identified as the author of this Work in accordance with the Copyright, Designs, and Patents Act 1988.

Information in this book is the opinion of the author, and is correct to the best of the author's knowledge, but is provided "as is" and without warranty to the maximum extent permissible under law. Content within this book is **not** intended as legal, tax, financial, medical, or any other form of professional advice.

While we have checked the content of this book carefully, in any educational book there is always the possibility of typographical errors, or other errors or omissions. We apologize if any such errors are found, and would appreciate if readers inform of any errors they might find, so we can update future editions/updates of this book.

IMPORTANT NOTICE: DO NOT ATTEMPT TO PERFORM ANY OF THE CHEMICAL REACTIONS DESCRIBED IN THIS BOOK. THE CHEMICAL REACTIONS DESCRIBED IN THIS BOOK ARE INCLUDED SOLELY FOR THE PURPOSE OF PROVIDING ILLUSTRATIVE EXAMPLES WHEN PERFORMING CALCULATIONS WITH PEN AND PAPER (AND HAVE NOT BEEN TESTED IN A LABORATORY). MANY CHEMICALS ARE TOXIC AND/OR OTHERWISE HAZARDOUS TO HUMAN HEALTH. MANY CHEMICAL REACTIONS ARE DANGEROUS, FOR EXAMPLE BY RELEASING DANGEROUS, FLAMMABLE OR TOXIC CHEMICALS, OR BY GENERATING LARGE AMOUNTS OF HEAT. THE QUANTITIES DESCRIBED IN THE EXAMPLES IN THIS BOOK MAY BE UNSUITABLE FOR LABORATORY EXPERIMENTS. CHEMICAL EXPERIMENTS OF ANY KIND SHOULD ONLY BE PERFORMED UNDER THE DIRECTION OF AN EXPERIENCED AND QUALIFIED CHEMIST, WITH APPROPRIATE SAFETY PRECAUTIONS, AND IN A SUITABLY EQUIPPED LABORATORY

Answers 2000 Limited is a private limited company registered in England under company number 3574155. Address and other information about information about Answers 2000 Limited can be found at www.ans2000.com

Updates, news & related resources from the author can be found at
http://www.suniltanna.com/moles

Information about other science books by the same author can be found at
http://www.suniltanna.com/science

Table of Contents

Introduction

Chapter 1: What is Stoichiometry & Key Definitions
Moles
Molar Mass & Relative Atomic Mass
Relative Molecular Mass & Relative Formula Mass
Mass Spectrometry
Questions
Answers to Chapter 1 Questions

Chapter 2: Mass Calculations
Converting Mass to Moles
Calculating the Mass Fraction and Percentage by Mass of Elements in a Compound
Calculating the Mole Fractions and Mole Percents of a Mixture
Calculating the Empirical Formula of a Compound
- Examples of Empirical Formulae
- Calculating an Empirical Formula from Masses of the Elements in a Sample
- Calculating an Empirical Formula from Percentage by Mass of the Elements

Calculating the Molecular Formula of a Compound
Calculating Masses in Reactions
Excess and Limiting Reactants
Questions
Answers to Chapter 2 Questions

Chapter 3: Solution Calculations
Volume and Concentration
Solutions Formulae
Converting Between Concentration Units
Titrations
- Titration Calculations using Moles Ratios
- Titration Calculations in One Step

Questions
Answers to Chapter 3 Questions

Chapter 4: Gas Calculations
Ideal Gases
Avogadro's Law
Ideal Gases at Room Temperature and Pressure
Ideal Gases at Standard Temperature and Pressure
Ideal Gas Law
Questions
Answers to Chapter 4 Questions

Chapter 5: Atom Economy
Calculating the Atom Economy of a Reaction
Calculating the Atom Economy of a Chemical Process Involving Multiple Steps
Improving Atom Economy
Questions
Answers to Chapter 5 Questions

Chapter 6: Chemical Yield
Absolute Yield and Molar Yield
Theoretical Yield
Actual Yield is Always Less Than Theoretical Yield
Fractional Yield and Percentage Yield
The Difference Between Atom Economy and Percentage Yield
How to Maximize Percentage Yield
Questions
Answers to Chapter 6 Questions

Chapter 7: Electrolysis Calculations
Moles of Electrons and the Faraday Constant
Special Considerations for Electroplating
Questions
Answers to Chapter 7 Questions

Chapter 8: Other Molar Constants & Units
Faraday Constant and Faradays
Einsteins
Kilogram-moles and Pound-moles

Chapter 9: Mole Factoids
Mole Day
10:23 Campaign

Conclusion

Introduction

For some time now, I have tutored both children and adults in science and math. This book is based on my own personal experience as a tutor, and is one of a series of books on different topics:

- If you want to find out about the other chemistry books that I have written, please visit http://www.suniltanna.com/chemistry

- If you want to find out about the other science books that I have written, please visit http://www.suniltanna.com/science

- For math books that I have written, visit: http://www.suniltanna.com/math

This book is about **stoichiometry** (calculating the amounts of different substances involved in a reaction or forming a compound) and **moles** (a mole is a measure of the amount of a substance). Don't worry if these words baffling right now – I will explain in much more terms what these terms mean in Chapter 1.

Throughout this book I will give plenty of examples, and each of the main chapters ends with questions (and answers). I strongly urge you to do the questions – as doing so will definitely improve your understanding.

Finally, at the very end of the book, I have included a couple of reference pages with the most important stoichiometry data & formulae and a periodic table of the elements.

IMPORTANT NOTICE: DO NOT ATTEMPT TO PERFORM ANY OF THE CHEMICAL REACTIONS DESCRIBED IN THIS BOOK. THE CHEMICAL REACTIONS DESCRIBED IN THIS BOOK ARE INCLUDED SOLELY FOR THE PURPOSE OF PROVIDING ILLUSTRATIVE EXAMPLES WHEN PERFORMING CALCULATIONS WITH PEN AND PAPER (AND HAVE NOT BEEN TESTED IN A LABORATORY). MANY CHEMICALS ARE TOXIC AND/OR OTHERWISE HAZARDOUS TO HUMAN HEALTH. MANY CHEMICAL REACTIONS ARE DANGEROUS, FOR EXAMPLE BY RELEASING DANGEROUS, FLAMMABLE OR TOXIC CHEMICALS, OR BY GENERATING LARGE AMOUNTS OF HEAT. THE QUANTITIES DESCRIBED IN THE EXAMPLES IN THIS BOOK MAY BE UNSUITABLE FOR LABORATORY EXPERIMENTS. CHEMICAL EXPERIMENTS OF ANY KIND SHOULD ONLY BE PERFORMED UNDER THE DIRECTION OF AN EXPERIENCED AND QUALIFIED CHEMIST, WITH APPROPRIATE SAFETY PRECAUTIONS, AND IN A SUITABLY EQUIPPED LABORATORY.

Chapter 1: What is Stoichiometry & Key Definitions

Stoichiometry is the study of the quantitative relationships and ratios of reactants and products in a chemical reaction. The term stoichiometry was coined by the German chemist Jeremias Benjaim Richter (March 10th, 1762 to April 14th, 1807) in 1792. The word derives from two Greek words; *stoicheion* (which means "element") and *metron* (which means "to measure").

Jeremias Benjaim Richter:

Let's consider an example... say, this chemical reaction:

$2H_2(g) + O_2(g) \rightarrow 2H_2O(g)$

You can see that 2 hydrogen (H_2) molecules react with 1 oxygen molecule (O_2) to produce 2 water (H_2O) molecules. In this case the ratio of hydrogen : oxygen : water molecules is 2 : 1 : 2. Moreover, with a little additional information (which we will learn in this book), it is possible to calculate the masses of different substances that react with each other, the mass of water that would be produced, and the volumes of gases.

Moles

In the above reaction, we saw the ratio of hydrogen, oxygen and water molecules in the reaction, but of course in real-life it is rarely if ever possible to count individual molecules. Fortunately, there are ways around this.

Because we know that 2 hydrogen molecules react with 1 oxygen molecule to produce 2 water molecules, it follows that:
- 2 *dozen* hydrogen molecules would react with 1 *dozen* oxygen molecules to produce 2 *dozen* water molecules.

- Likewise, 2 *thousand* hydrogen molecules would react with 1 *thousand* oxygen molecules to produce 2 *thousand* water molecules.
- Likewise, 2 *million* hydrogen molecules would react with 1 *million* oxygen molecules to produce 2 *million* water molecules.
- And so on...

Chemists use the term **mole** (symbol: **mol**) to refer a specific large number of atoms or molecules (we will come to discussing the precise number in a moment), so it follows that:
- 2 *moles* of hydrogen molecules would react with 1 *mole* of oxygen molecules to produce 2 *moles* of water molecules.
- Likewise, 0.2 *moles* of hydrogen molecules would react with 0.1 *moles* of oxygen molecules to produce 0.2 *moles* of water molecules.
- Likewise, 20 *moles* of hydrogen molecules would react with 10 *moles* of oxygen molecules to produce 20 *moles* of water molecules.
- And so on...

But how is this useful? It turns out that there are formulae to calculate exactly how many moles of a substance are in a given mass, volume of gas, volume of liquid at a particular concentration, and so forth. By using these formulae, one can calculate the amounts of substances that react together, and the amounts of products produced.

Today, the **mole** (symbol: **mol**) is part of the International System of Units, also known as the SI Unit system). But what is the definition of 1 mole?
- Prior to 1967, there were several definitions of 1 mole used at different times. This is mainly of historical interest now, so we will **not** discuss all these details. However, the important thing is that although the definitions varied, the actual number of particles in 1 mole was always similar (but **not** identical) to the present-day value of 1 mole.
- In 1967, a mole was defined as the number of atoms in 12 grams of carbon-12. The actual number of atoms (known as the **Avogadro number**) needed to be determined experimentally, but experiments in 2017 found it to be between $6.022140758 \times 10^{23}$ and $6.022140762 \times 10^{23}$.
- On November 16th, 2018 a new definition of mole was agreed by the General Conference on Weights and Measures (CGPM). The new definition, which officially takes effect on May 20th, 2019, defines 1 mole to be exactly $6.02214076 \times 10^{23}$, or written out as normal number, 602,214,076,000,000,000,000,000. This number is known as the **Avogadro constant** (symbol: N_A or L).

Both the Avogadro number and Avogadro constant are named after the Italian chemist, Amedeo Avogadro (August 9th, 1776 to July 9th, 1856), who did among other things, investigated the relationship between the volumes of gases and the number of particles that made up the gas (see Chapter 4).

Amedeo Avogadro:

Molar Mass & Relative Atomic Mass

The **molar mass** (symbol: **M**) of a substance is the mass per mole of particles of that substance. So, since 1 mole of carbon-12 weighs 12 grams, the molar mass of carbon-12 is 12 g mol^{-1} (which can also be written as 12 g/mol or as 0.012 kg mol^{-1} or as 0.012 kg/mol).

The **relative atomic mass** (symbol: A_r) of a substance is the **average** mass of an atom of a substance on a scale in which 1 atom of carbon-12 is exactly 12. Hence the relative atomic mass of carbon-12 is 12 – the same number (but without the units) as the molar mass in g mol^{-1}.

Note however that the relative atomic mass of the element carbon is **not** exactly 12. This is because carbon atoms are mixture of several different isotopes of different masses (in the case of carbon, the naturally occurring isotopes are carbon-12, carbon-13, and a very tiny amount of carbon-14), and hence the **average** mass of a carbon atom is **not** exactly 12. In fact, the average, taking into account the masses of each isotope and their relative abundances, turns out to be 12.0107.

Here is how the relative atomic mass of carbon is calculated:
- 98.9% of carbon is carbon-12, which has a relative isotopic mass of exactly 12.
- 1.1% of carbon is carbon-13, which has a relative isotopic mass of 13.0033548378.
- 0.0000000001% of carbon is carbon-14, which has a relative isotopic mass of 14.003241989.
- The relative atomic mass is therefore = (98. 9 × 12 + 1.1 × 13.0033548378 + 0.0000000001 × 14.003241989) ÷ 100 = 12.01 (4 significant figures).

Similarly, we can calculate the relative atomic mass of lithium:
- 7.59% of lithium is lithium-6, which has a relative isotopic mass of 6.015122795.

- 92.41% of lithium is lithium-7, which has a relative isotopic mass of 7.0160034366.
- The relative atomic mass is therefore = (7.59 × 6.015122795 + 92.41 × 7.0160034366) ÷ 100
  = 6.941 (4 significant figures).

Note: Of course, in all cases when calculating relative atomic mass, the value depends on the relative abundance of the various isotopes of the element concerned. The relative abundance of different isotopes varies depending on the source of the element – for example, a terrestrial sample of lithium will have a different mix of isotopes than one from deep space.

For all elements, we can look at the Periodic Table (a copy of which is included at the back of this book) to find their relative atomic mass. If you examine at the Periodic Table included with this book, you will see each individual box describing an element contains the relative atomic mass of that element. Of course, as relative atomic masses vary depending on the source of each element and the relative abundance of each isotope of that element, a Periodic Table intended for use on another planet or in another galaxy might contain slightly different relative atomic masses than one intended for use on Earth.

For Sodium (symbol: Na), the box looks like this, and the relative atomic mass is 22.98976:

The Periodic Table included with this book includes quite a lot of additional information, as well as several decimal places of precision in the relative atomic mass. Other Periodic Tables may omit some of this information or provide fewer decimal places. For example, a less detailed Periodic Table might simply tell you that Sodium has a relative atomic mass of 23, and an atomic number (the number of protons in an atom of the element) of 11.

Most of the time it is **not** necessary to include several decimal places of precision when using relative atomic masses in calculations, so most of the examples in this book will round the relative atomic masses to 1 decimal place (so the relative atomic mass of a sodium in a calculation will be taken as 23.0).

Relative Molecular Mass & Relative Formula Mass

Since we know the relative atomic masses of every element, we can also, using the same scale calculate the relative mass of any molecule. This is known as its relative molecular mass (symbol M_r) and is calculated by simply adding up the relative atomic masses of the atoms that make up the molecule – of course taking into account the number of atoms of each element. Moreover, in the case of a molecule, the molar mass of the molecule in g mol^{-1} is the same number as the relative molecular mass.

Example: We can calculate the relative molecular mass of water, H_2O:
- The molecule contains 2 hydrogen atoms and 1 oxygen atom.
- Looking at the Periodic Table, we see that hydrogen has a relative atomic mass of 1.00794, which we will round to 1.0.
- Likewise, we see that oxygen has a relative atomic mass of 15.9994, which we will round to 16.0.
- The relative molecular mass of water is thus (2 × 1.0) + (1 × 16.0) = 18.0.
- The molar mass of water is therefore 18.0 g mol^{-1}.

Example: We can calculate the relative molecular mass of ammonia, NH_3:
- The molecule contains a total of 1 nitrogen atom and 3 hydrogen atoms.
- Looking at the Periodic Table, we see that nitrogen has a relative atomic mass of 14.0067, which we round to 14.0.
- Hydrogen has a relative atomic mass of 1.00794, which we round to 1.0.
- The relative molecular mass of ammonia is thus (1 × 14.0) + (3 × 1.0) = 17.0.
- The molar mass of ammonia is therefore 17.0 g mol^{-1}.

Example: We can calculate the relative molecular mass of ethanol, CH_3CH_2OH:
- The molecule contains a total of 2 carbon atoms, 6 hydrogen atoms, and 1 oxygen atom.
- Looking at the Periodic Table, we see that carbon has a relative atomic mass of 12.0107, which we round to 12.0.
- Hydrogen has a relative atomic mass of 1.00794, which we round to 1.0.
- Oxygen has a relative atomic mass of 15.9994, which we round to 16.0.
- The relative molecular mass of ethanol is thus (2 × 12.0) + (6 × 1.0) + (1 × 16.0) = 46.0.
- The molar mass of ethanol is therefore 46.0 g mol^{-1}.

Ionic compounds and giant covalent structures do **not** form molecular units (instead they form large lattices), so it does **not** make sense to talk about relative molecular mass in such cases. However, we can still calculate the relative mass of a formula unit of the compound, and this is known as the relative formula mass (the symbol is also M_r). The relative formula mass is calculated by simply adding up the relative atomic masses of the atoms that make up the formula unit – of course taking into account the number of atoms of each element. This time the molar mass in g mol^{-1} is the same number as the relative formula mass.

Example: We can calculate the relative formula mass of calcium sulfate, $CaSO_4$:
- The formula unit contains 1 calcium atom, 1 sulfur atom and 4 oxygen atoms.
- Looking at the Periodic Table, we see that calcium has a relative atomic mass of 40.078, which we will round to 40.1.
- Likewise, we see that sulfur has a relative atomic mass of 32.065, which we round to 32.1.

- Again, oxygen has a relative atomic mass of 15.9994, which we round to 16.0.
- The relative formula of calcium sulfate is thus (1 × 40.1) + (1 × 32.1) + (4 × 16.0) = 136.2.
- The molar mass of calcium sulfate is therefore 136.2 g mol^{-1}.

Example: We can calculate the relative formula mass of indium (III) sulfate, $In_2(SO_4)_3$:
- The formula unit contains 2 indium atoms, 3 sulfur atoms and 12 oxygen atoms.
- Looking at the Periodic Table, we see that indium has a relative atomic mass of 114.818, which we will round to 114.8.
- Sulfur has a relative atomic mass of 32.065, which we round to 32.1.
- Oxygen has a relative atomic mass of 15.9994, which we round to 16.0.
- The relative formula of indium (III) sulfate is thus (2 × 114.8) + (3 × 32.1) + (12 × 16.0) = 517.9.
- The molar mass of indium (III) sulfate is therefore 517.9 g mol^{-1}.

Example: We can calculate the relative formula mass of hydrated copper (II) sulfate, $CuSO_4.5H_2O$:
- The formula unit contains 1 copper atom, 1 sulfur atom, 9 oxygen atoms and 10 hydrogen atoms.
- Looking at the Periodic Table, we see that copper has a relative atomic mass of 63.546, which we will round to 63.5.
- Sulfur has a relative atomic mass of 32.065, which we round to 32.1.
- Oxygen has a relative atomic mass of 15.9994, which we round to 16.0.
- Hydrogen has a relative atomic mass of 1.00794, which we round to 1.0.
- The relative formula of hydrated copper (II) sulfate is thus (1 × 63.5) + (1 × 32.1) + (9 × 16.0) + (10 × 1.0) = 249.6.
- The molar mass of hydrated copper (II) sulphate is therefore 249.6 g mol^{-1}.

Mass Spectrometry

A **mass spectrometer** is a device used when analyzing chemical samples. It operates by ionizing particles of the sample and measuring the **mass-to-charge ratio** (often abbreviated m/Q or m/z) of the ions produced. Since almost of the ions will have lost a single electron, and therefore will have the exact same charge, it is also possible to determine the relative mass of each type of ion, as well as the relative abundance of each of them.

When the sample contains a pure compound, the mass spectrometer will generally detect an ionized version of the entire molecule (known as the **molecular ion**) as well as ionized fragments of the molecule.
- From the m/Q of the molecular ion, we can determine the relative mass of a molecule of the original compound.
- Although we will **not** discuss it in this book, the ion fragments can be used to help distinguish between **isomers** – that is compounds that have an identical molecular-formulae but different special arrangements of the atoms.

Mass spectrometer:

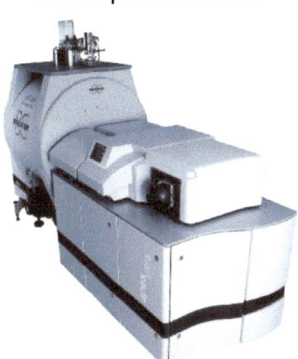

For example, if placing butane ($CH_3CH_2CH_2CH_3$) in a mass spectrometer, you would probably detect ions with m/Q of 58.0, 43.0, 29.0, 15.0 and 1.0. These would correspond to:
- 58.0 would correspond to the molecular ion $CH_3CH_2CH_2CH_3^+$, which has a relative mass of (4 × 12.0) + (10 × 1.0) = 58.0.
- 43.0 would correspond to the fragment ion $CH_3CH_2CH_2^+$, which has a relative mass of (3 × 12.0) + (7 × 1.0) = 43.0.
- 29.0 would correspond to the fragment ion $CH_3CH_2^+$, which has a relative mass of (2 × 12.0) + (5 × 1.0) = 29.0.
- 15.0 would correspond to the fragment ion CH_3^+, which has a relative mass of (1 × 12.0) + (3 × 1.0) = 15.0.
- 1.0 would correspond to the fragment ion H^+, which has a relative mass of 1.0.

One important thing to note however as that the mass spectrometer measures the m/Q of individual ions (and each ion contains atoms of a particular isotope), whereas the Periodic Table shows the **average** relative atomic mass of each element (and hence allows you to calculate the **average** relative molecular mass of compounds).

For example, consider methyl chloride CH_3Cl:
- If we calculate the relative molecular mass of methyl chloride using the Periodic Table, we get 50.5. This is because (1 × 12.0) + (3 × 1.0) + (1 × 35.5) = 50.5.
- However, if we were to use a mass spectrometer, we would see molecular ions with m/Q of 50 and 52. This is because chlorine has two common isotopes, chlorine-35 (^{35}Cl) with A_r of 35.0, and chlorine-37 (^{37}Cl) with A_r of 37.0. Hence the actual mass of the molecular ion can either be (1 × 12.0) + (3 × 1.0) + (1 × 35.0) = 50.0 or (1 × 12.0) + (3 × 1.0) + (1 × 37.0) = 52.0. Note: we do **not** need to worry about other isotopes of carbon or hydrogen, as although present, they only represent a very tiny fraction of the carbon and hydrogen.

I mentioned previously that usually in this book we will round relative atomic and molecular masses to one decimal place. However, mass spectrometry is one case where we might want to use more precise relative atomic masses – although we will want to use the precise relative atomic masses of isotopes (relative isotopic masses) rather than the **average** relative atomic masses found in the Periodic Table. This is because by using precise masses, we can distinguish between compounds which would otherwise appear to have identical relative molecular masses.

First, here are the precise relative atomic masses of the principle isotopes of hydrogen, carbon and oxygen:

Isotope	Precise Relative Atomic Mass
^1H	1.007825032241
^{12}C	12.000000000000
^{16}O	15.99491461956

Now, consider butane (C_4H_{10}) and propanal (C_2H_5CHO):
- Using the relative atomic masses of the elements to one decimal place, the relative molecular mass of butane is (4 × 12.0) + (10 × 1.0) = 58.0.
- Likewise, using the relative atomic masses of the elements to one decimal place, the relative molecular mass of propanal is (3 × 12.0) + (6 × 1.0) + (1 × 16.0) = 58.0.
- Using precise relative isotopic masses, the m/Q of the molecular ion of butane is (4 × 12.000000000000) + (10 × 1.007825032241) ≈ 58.07825032.
- Using precise relative isotopic masses, the m/Q of the molecular ion of propanal is (3 × 12.000000000000) + (6 × 1.007825032241) + (1 × 15.99491461956) ≈ 58.04186481.
- If you had a mass spectrometer which provided at least three significant figures of accuracy in m/Q values, you would be able to distinguish between the molecular ions of butane and propanal. In fact, if your mass spectrometer gave a few more significant figures, you would be able to identify the molecular formula for **any** molecular ion just from its m/Q, since only one particular combination of atoms would match it to sufficient accuracy.

Questions

1. If I have 7 moles of carbon dioxide (CO_2). How many moles of carbon atoms are there?
2. If I have 7 moles of carbon dioxide (CO_2). How many moles of oxygen atoms are there?
3. How many molecules are in in 0.3 moles of ammonia (NH_3)?
4. How many moles of nitrogen atoms are in 0.3 moles of ammonia (NH_3)?
5. How many nitrogen atoms are in 0.3 moles of ammonia (NH_3)?
6. How many moles of hydrogen atoms are in 0.3 moles of ammonia (NH_3)?
7. How many hydrogen atoms are in 0.3 moles of ammonia (NH_3)?
8. What is the relative atomic mass of silicon (Si)?
9. What is the molar mass of silicon (Si)?
10. What is the relative molecular mass of carbon dioxide (CO_2)?
11. What is the molar mass of carbon dioxide (CO_2)?
12. What is the relative molecular mass of pentane (C_5H_{12})?

13. Chlorine consists of two isotopes: 75.78% chlorine-35 of isotopic mass 34.96885268, and 24.22% of chlorine-37 of isotopic mass 36.96590259. What is the relative atomic mass of chlorine?
14. What is the M_r of phenol (C_6H_5OH)?
15. What is relative molecular mass of aniline ($C_6H_5NH_2$)?
16. How many moles of chromium atoms are in 1 mole of hydrated chromium (III) sulfate ($Cr_2(SO_4)_3.18H_2O$)?
17. How many moles of sulfur atoms are in 1 mole of hydrated chromium (III) sulfate ($Cr_2(SO_4)_3.18H_2O$)?
18. How many moles of oxygen atoms are in 1 mole of hydrated chromium (III) sulfate ($Cr_2(SO_4)_3.18H_2O$)?
19. How many moles of hydrogen atoms are in 1 mole of hydrated chromium (III) sulfate ($Cr_2(SO_4)_3.18H_2O$)?
20. What is relative formula mass of hydrated chromium (III) sulfate ($Cr_2(SO_4)_3.18H_2O$)?
21. What is the M_r of sodium carbonate (Na_2CO_3)?
22. How many moles of chromium atoms are in 1 mole of chromium (III) oxide (Cr_2O_3)?
23. How many moles of oxygen atoms are in 1 mole of chromium (III) oxide (Cr_2O_3)?
24. What is the molar mass of methane (CH_4)?
25. What is the relative formula mass of potassium permanganate ($KMnO_4$)?
26. What is the molar mass of glucose ($C_6H_{12}O_6$)?
27. How many moles of carbon atoms are in 0.5 moles of heptane (C_7H_{16})?
28. How many moles of hydrogen atoms are in 0.5 moles of heptane (C_7H_{16})?
29. How many moles of nitrogen atoms are there 1 mole of ammonium sulfate (($NH_4)_2SO_4$)?
30. How many moles of hydrogen atoms are there 1 mole of ammonium sulfate (($NH_4)_2SO_4$)?
31. How many moles of sulfur atoms are there 1 mole of ammonium sulfate (($NH_4)_2SO_4$)?
32. How many moles of oxygen atoms are there 1 mole of ammonium sulfate (($NH_4)_2SO_4$)?
33. What is the relative formula mass of ammonium sulfate (($NH_4)_2SO_4$)?
34. What is the relative atomic mass of oxygen?
35. What is the A_r of oxygen?
36. What is the relative molecular mass of oxygen (O_2)?
37. What is the M_r of oxygen?
38. The relative atomic mass of lithium-7 (^7Li) is 7.01600344. However, the Periodic Table gives the relative atomic mass of lithium as being 6.941. Why are the two figures so different?
39. An alkane has an M_r of 58.0. Is it methane (CH_4), ethane (C_2H_6), propane (C_3H_8), butane (C_4H_{10}) or pentane (C_5H_{12})?
40. A student looks at the read-out from a mass spectrometer and says that the m/Q of a compound is 44.0. Why might it be difficult to tell whether the compound is carbon dioxide (CO_2), propane (C_3H_8) or ethanal (CH_3CHO)? How could she tell which compound it is?

Answers to Chapter 1 Questions

1. 7 because each molecule of carbon dioxide contains 1 carbon atom.
2. 14 because each molecule of carbon dioxide contains 2 oxygen atoms.
3. 0.3 moles = 0.3 × 6.02214076 × 10^{23} = 1.806642228 × 10^{23} molecules.
4. 0.3 because each molecule of ammonia contains 1 nitrogen atom.
5. 0.3 moles = 0.3 × 6.02214076 × 10^{23} = 1.806642228 × 10^{23} atoms.
6. 0.9 because each molecule of ammonia contains 3 hydrogen atoms, and 3 × 0.3 = 0.9.
7. 0.9 moles = 0.9 × 6.02214076 × 10^{23} = 5.419926684 × 10^{23} atoms.
8. 28.0855.
9. 28.0855 g mol^{-1}.
10. (1 × 12.0) + (2 × 16.0) = 44.0.
11. 44.0 g mol^{-1}.
12. (5 × 12.0) + (12 × 1.0) = 72.0.
13. (75.78 × 34.96885268 + 24.22 × 36.96590259) ÷ 100 = 35.45253817 = 35.45 (4 significant figures).
14. (6 × 12.0) + (6 × 1.0) + (1 × 16. 0) = 94.0.
15. (6 × 12.0) + (7 × 1.0) + (1 × 14. 0) = 93.0.
16. 2 moles.
17. 3 moles.
18. 30 moles.
19. 36 moles.
20. (2 × 52.0) + (3 × 32.1) + (12 × 16.0) + (36 × 1.0) + (18 × 16.0) = 716.3.
21. (2 × 23.0) + (1 × 12.0) + (3 × 16.0) = 106.0.
22. 2 moles.
23. 3 moles.
24. (1 × 12.0) + (4 × 1.0) = 16.0.
25. (1 × 39.1) + (1 × 54.9) + (4 × 16.0) = 155.0.
26. (6 × 12.0) + (12 × 1.0) + (6 × 16.0) = 180.0.
27. 7 × 0.5 = 3.5.
28. 16 × 0.5 = 8.
29. 2 moles.
30. 8 moles.
31. 1 mole.
32. 4 moles.
33. (2 × 14.0) + (8 × 1.0) + (1 × 32.1) + (4 × 16.0) = 132.1.
34. 15.9994.
35. 15.9994.
36. (2 × 16.0) = 32.0.
37. Because the question says M_r rather than A_r, we can infer the question is talking about oxygen molecules (O_2) rather than oxygen atoms (O). Hence, (2 × 16.0) = 32.0.

38. Because there are several isotopes of lithium each with a different relative atomic mass. The Periodic Table contains the average relative atomic mass of a lithium atom.

39. We can calculate the M_r of each alkane:
- The M_r of methane is $(1 \times 12.0) + (4 \times 1.0) = 16.0$.
- The M_r of ethane is $(2 \times 12.0) + (6 \times 1.0) = 30.0$.
- The M_r of propane is $(3 \times 12.0) + (8 \times 1.0) = 44.0$.
- The M_r of butane is $(4 \times 12.0) + (10 \times 1.0) = 58.0$.
- The M_r of pentane is $(5 \times 12.0) + (12 \times 1.0) = 72.0$.

Hence the alkane must be butane as it is the only one with a matching M_r.

40. Using the precise relative atomic masses of hydrogen-1, carbon-12 and oxygen-16:
- The m/Q for carbon dioxide would be $(1 \times 12.000000000000) + (2 \times 15.99491461956) = 43.98982924$ which rounded to three significant figures is 44.0.
- The m/Q for propane would be $(3 \times 12.000000000000) + (8 \times 1.007825032241) = 44.06260026$ which rounded to 3 significant figures is 44.1
- The m/Q for ethanal would be $(2 \times 12.000000000000) + (4 \times 1.007825032241) + (1 \times 15.99491461956) = 44.02621475$ which rounded to 3 significant figures is 44.0.

Since, the m/Q reading has only 3 significant figures, we can **not** distinguish between carbon dioxide and ethanal. However, if we had a more precise m/Q reading (in this case 4 significant figures would be enough), we would be able to. Alternatively, we might get additional clues as to the identity of the compound by looking at the m/Q values of fragment ions, or by performing chemical tests on the sample.

Chapter 2: Mass Calculations

In this chapter, we will discuss the relationship between the mass of a substance and the number of particles in the substance.

Converting Mass to Moles

In Chapter 1, we established that the molar mass, which is the mass in grams of 1 mole of a substance, is the same as the substance's A_r (relative atomic mass) or M_r (relative molecular mass or relative formula mass). Clearly this means that 2 moles of a substance, that is twice as many particles, will have double the mass, 3 moles would have three times the mass, and so on.

For example, since the relative atomic mass of antimony (Sb) is 121.8, it follows that:
- 1 mole of antimony atoms would have a mass of 121.8 g.
- 2 moles of antimony atoms would have a mass of 2 × 121.8 = 243.6 g.
- 3 moles of antimony atoms would have a mass of 3 × 121.8 = 365.4 g.
- 7 moles of antimony atoms would have a mass of 7 × 121.8 = 852.6 g.
- 0.5 moles of antimony atoms would have a mass of 0.5 × 121.8 = 60.9 g.
- And so on…

Likewise, since the relative molecular mass of ethane (C_2H_6) is 30.0, it follows that:
- 1 mole of ethane molecules would have a mass of 30.0 g.
- 2 moles of ethane molecules would have a mass of 2 × 30.0 = 60.0 g.
- 3 moles of ethane molecules would have a mass of 3 × 30.0 = 90.0 g.
- 0.5 moles of ethane molecules would have a mass of 0.5 × 30.0 = 15.0 g.
- 0.93 moles of ethane molecules would have a mass of 0.93 × 30.0 = 27.9 g.
- And so on…

In general:
- **Mass of atoms (grams) = Moles of atoms × A_r**
- **Mass of molecules (grams) = Moles of molecules × M_r**
- **Mass of formula units (grams) = Moles of formula units × M_r**

We can also do these types of calculations the other way around too. Simply rearranging the above formulae, we see that dividing the mass in grams of substance by its A_r or M_r would give us the number of moles. In general, the formulae for calculating moles from mass are:
- **Moles of atoms = Mass (grams) ÷ A_r**
- **Moles of molecules = Mass (grams) ÷ M_r**
- **Moles of formula units = Mass (grams) ÷ M_r**

So, applying these formulae:
- The A_r of gold (Au) is 197.0. Therefore a 250-gram bar of gold contains 250 ÷ 197.0 = 1.27 moles of gold atoms.
- The A_r of silver (Ag) is 107.9. Therefore 3 moles of silver have a mass of 3 × 107.9 = 323.7 g.
- The M_r of water (H_2O) is 18.0. Therefore 900 grams of water contains 900 ÷ 18.0 = 50 moles of water molecules.
- The Mr of propane (C_3H_8) is 44.0. Therefore 25 moles of propane have a mass of 25 × 44.0 = 1,100 g = 1.100 kg.
- The M_r of quartz (SiO_2) is 60.1. The largest cluster of quartz crystals currently on display has a mass of 14,100 kg = 14,100,000 g. The cluster therefore contains 14,100,000 ÷ 60.1 = 234,609 moles of formula units of SiO_2.
- The M_r of chromium (III) oxide (Cr_2O_3) is 100.0. Therefore 7 moles of chromium (III) oxide have a mass 7 × 100.0 = 700 g.

Chromium (III) oxide:
(number of moles present undetermined)

Calculating the Mass Fraction and Percentage by Mass of Elements in a Compound
You can calculate the mass fraction (between 0 and 1) by mass of each element in a compound by adding up the relative atomic masses of all atoms of that element, and then dividing by the M_r of the compound. To convert the mass fraction into a percentage (known as the percentage by mass or mass percentage), simply multiply the mass fraction by 100.

Example: What is the percentage by mass of each of the elements in potassium permanganate ($KMnO_4$)?
- M_r = (1 × 39.1) + (1 × 54.9) + (4 × 16.0) = 158.0.
- The mass fraction of potassium in potassium permanganate is (1 × 39.1) ÷ 158.0 = 0.247 (3 significant figures).
- The percentage by mass of potassium in potassium permanganate is 0.252 × 100 = 24.7% (3 significant figures).
- The mass fraction of manganese in potassium permanganate is (1 × 54.9) ÷ 158.0 = 0.347 (3 significant figures).

- The percentage by mass of manganese in potassium permanganate is 0.347 × 100 = 34.7% (3 significant figures).
- The mass fraction of oxygen in potassium permanganate is (4 × 16.0) ÷ 158.0 = 0.405 (3 significant figures).
- The percentage by mass of oxygen in potassium permanganate is 0.405 × 100 = 40.5% (3 significant figures).

Potassium permanganate:
(contains 24.7% potassium, 34.7% manganese, 40.5% oxygen)

Calculating the Mole Fractions and Mole Percents of a Mixture

The mole fraction (also known as molar fraction or number fraction or amount fraction or amount-of-substance fraction, symbol: x or χ, or y when talking about gases) is the amount of a constituent of a mixture, measured in moles, divided by the total amount of the mixture, also measured in moles. Multiplying the mole fraction by 100 gives the mole percent (also known as molar percentage or molar proportion, symbol: mol%).

It should be noted that although calculations of the mole fraction and mole percent are usually done using the number of moles of the substances, they are both dimensionless quantities and do **not** really have much inherently to do with moles. They simply express the proportion of all particles in a mixture which are of a particular type. It is for this reason that some authorities encourage the use of names that don't include the word "mole", such as amount fraction or amount-of-substance fraction.

Example: A mixture contains 3 moles of oxygen (O_2) and 12 moles of nitrogen (N_2). What is the mole fraction and mole percent of each of the gases?
- Total moles = 3 + 12 = 15.
- Mole fraction O_2 = 3 ÷ 15 = 0.2.
- Mole percent O_2 = 0.2 × 100 = 20%.
- Mole fraction N_2 = 12 ÷ 15 = 0.8.
- Mole percent N_2 = 0.8 × 100 = 80%.

Calculating the Empirical Formula of a Compound

The **empirical formula** of a compound is the lowest whole number ratio between constituent atoms of the compound.

Examples of Empirical Formulae

Let us begin by looking at some compounds and reviewing the empirical formula of each one.

Example: What is the empirical formula of ethane (C_2H_6)?
- The **molecular formula** of ethane is C_2H_6 since one molecule of ethane contains 2 carbon and 6 hydrogen atoms.
- The ratio of carbon to hydrogen atoms is 2 : 6 which can be reduced to 1 : 3. The empirical formula of ethane is therefore CH_3.

Example: What is the empirical formula of ethyl benzene (C_8H_{10})?
- The molecular formula of ethyl benzene is C_8H_{10} since one molecule of ethyl benzene contains 8 carbon and 10 hydrogen atoms.
- The ratio of carbon to hydrogen atoms is 8 : 10 which can be reduced to 4 : 5. The empirical formula of ethyl benzene is therefore C_4H_5.

Example: What is the empirical formula of cyclopentane (C_5H_{10})?
- The molecular formula of cyclopentane is C_5H_{10} since one molecule of cyclopentane contains 5 carbon and 10 hydrogen atoms.
- The ratio of carbon to hydrogen atoms is 5 : 10 which can be reduced to 1 : 2. The empirical formula of cyclopentane is therefore CH_2.

However:
- The molecular formula of propane is C_3H_8 since one molecule of propane contains 3 carbon and 8 hydrogen atoms.
- The ratio of carbon to hydrogen atoms is 3 : 8 which can **not** be reduced. The empirical formula of propane is therefore C_3H_8.

Calculating an Empirical Formula from Masses of the Elements in a Sample

If we know the masses of the constituent elements of a compound, it is possible to work out the empirical formula of that compound by examining the ratios between the number of moles of each element present. The steps are:
- **Step 1:** Calculate the number of moles of each element present by dividing the mass of that element present by its A_r.
- **Step 2:** Produce a ratio between the number of atoms of each element by dividing the number of moles of each element by the number of moles of the least abundant element.

- **Step 3:** Allowing for rounding errors, generate a whole number ratio between the elements, and thus an empirical formula. Note: In some cases you may find a ratio such as 1 : 1.5 which is **not** close to a whole number ratio – if this type of situation occurs, you will need to double, triple, etc., ratio to get a whole number ratio, so 1 : 1.5 should be doubled to 2 : 3.

Let's try an example...

Example: We wish to find the empirical formula of an organic chemical. An 80 g sample of an organic chemical was analyzed and was found to contain 43.6 g of carbon, 7.3 g of hydrogen and 29.1 g of oxygen.

We begin by looking up the A_r of each element present (12.0, 1.0 and 16.0 respectively), and then proceed through the steps, calculating the moles of each element present (step 1), the ratio between the moles (step 2), and rounding this to a whole number ratio (step 3).

Setting out our work in a table, would produce something like this:

	Carbon	Hydrogen	Oxygen
Mass (g)	43.6	7.3	29.1
A_r	12.0	1.0	16.0
Step 1: Moles = Mass ÷ A_r	43.6 ÷ 12.0 = 3.63	7.3 ÷ 1.0 = 7.3	29.1 ÷ 16.0 = 1.82
Step 2: Ratio	3.63 ÷ 1.82 = 1.99	7.3 ÷ 1.82 = 4.01	1.82 ÷ 1.82 = 1
Step 3: Whole number ratio	2	4	1

Since the ratio of carbon : hydrogen : oxygen is 2 : 4 : 1, the empirical formula of this compound is therefore C_2H_4O.

Calculating an Empirical Formula from Percentage by Mass of the Elements

In some cases, instead of getting the mass in grams of each element in a sample of compound, you might be given the mass fraction or the percentage by mass of each element. In both cases, you can simply say calculate on the basis of what a sample of 100 g would contain, by dividing up the 100 g according to the fractions or percentages:
- If you are given mass fractions, multiply each mass fraction by 100.
- If you are given percentages by mass, the amount in grams is the same number as the percentage.

For example, if a sample contained 50% element X by mass, 35% element Y by mass, and 15% element Z by mass, you could simply do your calculations assuming 50 g of element X, 35 g of element Y, and 15 g of element Z.

Let's try an actual example...

Example: A sample of an unknown inorganic chemical is analyzed. It is found to be 81% calcium by mass, and 19% nitrogen. What is the empirical formula of the compound?

If we had a 100 g sample of the compound – such a sample would contain 81 g of calcium and 19 g of nitrogen. Then we proceed as before: We look up the A_r of each element present (40.1 and 14.0 respectively), and then proceed through the steps, calculating the moles of each element present (step 1), the ratio between the moles (step 2), and working to get a whole number ratio (step 3).

In this particular example, you will notice that the initial ratio that we reached in step 3 was 1.5 : 1, so we double it to reach the final whole number ratio of 3 : 2.

	Calcium	Nitrogen
% of Total Mass	81	19
Mass (g) if 100g	81	19
A_r	40.1	14.0
Step 1: Moles = Mass ÷ A_r	81 ÷ 40.1 = 2.02	19 ÷ 14.0 = 1.36
Step 2: Ratio	2.02 ÷ 1.36 = 1.49	1.36 ÷ 1.36 = 1
Step 3: Rounded ratio	1.5	1
Step 3: Whole number ratio	1.5 × 2 = 3	1 × 2 = 2

Since the ratio of calcium : nitrogen is 3 : 2, the empirical formula of this compound is therefore Ca_3N_2.

Calculating the Molecular Formula of a Compound

If you **only** know the empirical formula of a compound, it is **not** possible to determine the molecular formula. This is because the molecular formula could be any whole number multiple of the empirical formula.

However, if you know **both** the empirical formula **and** the relative molecular mass (M_r) of the compound, it is possible to determine the molecular formula. All you have to do is find which whole number multiple of the empirical formula will have the correct relative molecular mass.

So, let's try an example of this:

Example: A sample of an unknown organic compound contains 83% carbon and 17% hydrogen by mass. A mass spectrometer determines that the m/Q of the molecular ion is 58.0. What is the empirical formula of the compound? What is the molecular formula?

First, we calculate the empirical formula – using the method previously described:

If we had a 100 g sample of the compound – such a sample would contain 83 g of carbon and 17 g of hydrogen. We look up the A_r of each element present (12.0 and 1.0 respectively), and then proceed through the steps, calculating the moles of each element present (step 1), the ratio between the moles (step 2), and working to get a whole number ratio (step 3). In this particular example, you will notice that the initial ratio we reached in step 3 was 1 : 2.5, so we double that to reach the final whole number ratio of 2 : 5. The empirical formula is thus C_2H_5.

	Carbon	Hydrogen
% of Total Mass	83	17
Mass (g) if 100g	83	17
A_r	12.0	1.0
Step 1: Moles = Mass ÷ A_r	83 ÷ 12.0 = 6.92	17 ÷ 1.0 = 17.00
Step 2: Ratio	6.92 ÷ 6.92 = 1	17.00 ÷ 6.92 = 2.46
Step 3: Rounded ratio	1	2.5
Step 3: Whole number ratio	1 × 2 = 2	2.5 × 2 = 5

Once we know the empirical formula, it is easy to determine the molecular formula: The molecular formula of the compound must be a whole number multiple of the empirical formula, so:
- If the multiple was × 1 then the molecular formula would be C_2H_5 and the M_r would be (2 × 12.0) + (5 × 1.0) = 29.0.
- If the multiple was × 1 then the molecular formula would be C_4H_{10} and the M_r would be (4 × 12.0) + (10 × 1.0) = 58.0.
- If the multiple was × 1 then the molecular formula would be C_6H_{15} and the M_r would be (6 × 12.0) + (15 × 1.0) = 87.0.
- And so on…

We know that the m/Q of the molecular ion is 58.0, so it follows that the M_r of the compound must also be 58.0. The molecular formula of the compound must therefore be C_4H_{10} as this is the only option with the correct M_r.

Calculating Masses in Reactions
A common requirement in chemistry is to calculate the masses of reactants or products in a chemical equation. There are two pieces of information that you can use to help you:
- A chemical equation gives the ratio between the number of moles of each of the reactants and each of the products. Therefore, if you know the number of moles of any one reactant

or product, you can use the ratio to determine the number of moles of each of the other reactants or products.
- The **law of conservation of mass** means that the total mass of products must always equal the total mass of the reactants.

Example: Let us consider heating of 12 g of calcium carbonate ($CaCO_3$) using a Bunsen burner. The calcium carbonate undergoes thermal decomposition to produce calcium oxide (CaO) and carbon dioxide (CO_2) according to the following chemical equation:

$CaCO_3(s) \rightarrow CaO(s) + CO_2(g)$

Since none of the species in the chemical equation have a large number written in front of them, they each should be considered to have an implicit 1 in front of them. Therefore, the ratio of $CaCO_3$: CaO : CO_2 is 1 : 1 : 1.

We can also calculate the M_r of each species:
- M_r $CaCO_3$ = (1 × 40.1) + (1 × 12.0) + (3 × 16.0) = 100.1.
- M_r CaO = (1 × 40.1) + (1 × 16.0) = 56.1.
- M_r CO_2 = (1 × 12.0) + (2 × 16.0) = 44.0.

Using Moles = Mass ÷ M_r, we can calculate:
- Moles $CaCO_3$ = 12 ÷ 100.1 = 0.11988.

Since the ratio of $CaCO_3$: CaO : CO_2 is 1 : 1 : 1, we also know:
- Moles CaO = Moles of $CaCO_3$ = 0.11988.
- Moles CO_2 = Moles of $CaCO_3$ = 0.11988.

Then using Mass = Moles × M_r, we can calculate the mass of each product:
- Mass CaO = 0.11988 × 56.1 = 6.725268 g ≈ 6.73 g (3 significant figures).
- Mass CO_2 = 0.11988 × 44.0 = 5.27472 g ≈ 5.27 g (3 significant figures).

You will notice that the mass of the reactants is 12 g, and the total mass of the products is 6.73 + 5.27 = 12 g too. Of course, as described above, the total mass of the products will always equal the total mass of reactants for any chemical reaction, because of the law of conservation of mass.

Example: For our second example let us consider the combustion of hydrogen (H_2) in oxygen (O_2) to produce water (H_2O):

$2H_2(g) + O_2(g) \rightarrow 2H_2O(g)$

Let us say in this example, we wish to burn 0.5 g of hydrogen, and our questions are:
- What mass of oxygen is required for the hydrogen undergo complete combustion?
- What mass of water is produced by the reaction?

The first thing that you should realize from examining the chemical equation is that the ratio of $H_2 : O_2 : H_2O$ is 2 : 1 : 2.

We can also calculate the M_r of each species:
- $M_r\ H_2 = (2 \times 1.0) = 2.0$.
- $M_r\ O_2 = (2 \times 16.0) = 32.0$.
- $M_r\ H_2O = (2 \times 1.0) + (1 \times 16.0) = 18.0$.

Using Moles = Mass ÷ M_r, we can calculate:
- Moles H_2 = 0.5 ÷ 2.0 = 0.25.

Since the ratio of $H_2 : O_2 : H_2O$ is 2 : 1 : 2, we also know:
- Moles O_2 = Moles H_2 ÷ 2 = 0.25 ÷ 2 = 0.125.
- Moles H_2O = Moles H_2 = 0.25.

Then using Mass = Moles × M_r, we can calculate the mass of oxygen required:
- Mass O_2 = 0.125 × 32.0 = 4.0 g.

As far as calculating the mass of water produced is concerned, there are two different ways to do it. We can either use Mass = Moles × M_r, or we can use the law of conservation of mass:
- Mass H_2O = Moles × M_r = 0.25 × 18.0 = 4.5 g.
- Law of conservation of mass: Mass of products (H_2O) = Total mass of reactants (H_2 and O_2) = 0.5 + 4.0 = 4.5 g.

Excess and Limiting Reactants

So far, we have reviewed reactions where there is exactly the right amount of each reactant for it to react completely. Very often however you might find yourself in a situation where there is too much of one reactant (it is said to be in excess or simply called the excess reacatant) and too little of another (it is called the limiting reactant).

In such cases, the limiting reactant present will react completely with the appropriate amount of the excess reactant, but some of the excess reactant will remain unreacted.

Let's try an example...

You Can Do Chemistry: Moles & Stoichiometry

Example: Imagine dropping a 10 g piece of sodium (Na) into a large bucket contain 5000 g of water (H_2O). **Note: Do not try this!**

The following reaction would occur:

$2Na(s) + 2H_2O(l) \rightarrow 2NaOH(aq) + H_2(g)$

The questions are:
- Which of the reactants is the excess reactant, and which is the limiting reactant?
- How many moles of sodium (Na) and water (H_2O) will actually react?
- What masses of sodium (Na) and water (H_2O) will actually react?
- How many moles of sodium hydroxide (NaOH) and hydrogen (H_2) will be produced by the reaction?
- What masses of sodium hydroxide (NaOH) and hydrogen (H_2) will be produced by the reaction?

The first thing that you should realize from examining the chemical equation is that the ratio of Na : H_2O : NaOH : H_2 is 2 : 2 : 2 : 1.

We can also calculate the M_r of each species:
- M_r Na = (1 × 23.0) = 23.0.
- M_r H_2O = (2 × 1.0) + (1 × 16.0) = 18.0.
- M_r NaOH = (1 × 23.0) + (1 × 16.0) + (1 × 1.0) = 40.0.
- M_r H_2 = (2 × 1.0) = 2.0.

Using Moles = Mass ÷ M_r, we can calculate:
- Moles Na = 10 ÷ 40.0 = 0.25.
- Moles H_2O = 5000 ÷ 18.0 = 277.78.

Since the ratio Na : H_2O is 2 : 2, we know:
- The 0.25 moles of Na that are present, could react with 0.25 moles of H_2O. There is more than enough H_2O present. So, H_2O is the excess reactant.
- The 277.78 moles of H_2O that are present, could react with 277.78 moles of Na. However, there is **not** enough Na present, and only enough Na for 0.25 moles of water to react with. So, Na is the limiting reactant.

Now that we know that 0.25 moles of Na will actually react, we can use the 2 : 2 : 2 : 1 ratio of Na : H_2O : NaOH : H_2 to conclude:
- 0.25 moles of Na will react.
- 0.25 moles of H_2O will react.

- 0.25 moles of NaOH will be produced.
- 0.25 ÷ 2 = 0.125 moles of H_2 will be produced.

Then using Mass = Moles × M_r, we can calculate the mass of each reactant that actually reacted, and the mass of each product produced:
- Mass of Na that reacts = 0.25 × 23.0 = 5.75 g.
- Mass of H_2O that reacts = 0.25 × 18.0 = 4.5 g.
- Mass of NaOH produced = 0.25 × 40.0 = 10 g.
- Mass of H_2 produced = 0.125 × 2 = 0.25 g.

Please notice that:
- As water (H_2O) was vastly in excess, only a small proportion of the water available reacted: 0.25 moles reacted of out 277.78 moles available, or to put it another way, 4.5 g of water reacted out of the 5000 g of water available in the bucket.
- If you compare the total mass of reactants that reacted (Na and H_2O) with the total mass of the products produced (NaOH and H_2), you will find that they are the same (because of the law of conservation of mass). In this experiment, the mass of Na and H_2O that reacted = 5.75 + 4.5 = 10.25 g, and the mass of the products produced = 10 + 0.25 = 10.25 g.

Questions
1. How many moles of water (H_2O) are present in a bottle containing 500 g?
2. How many moles of iron (III) oxide (Fe_2O_3) are present in a 500 g sample?

Iron (III) oxide:

3. How many moles of iron atoms are present in 500 g sample of iron (III) oxide (Fe_2O_3)?
4. How many moles of oxygen atoms are present in 500 g sample of iron (III) oxide (Fe_2O_3)?
5. What is the mass of 0.02 moles of silver chloride (AgCl)?
6. What is the mass of 1.87 moles of potassium permanganate ($KMnO_4$)?

7. What is the percentage by mass of each element in sodium carbonate (Na_2CO_3)?
8. What is the percentage by mass of each element in silver chloride (AgCl)?
9. What is the percentage by mass of each element in nitride (Ca_3N_2)?
10. What is the percentage by mass of each element in ammonium nitrate (NH_4NO_3)?
11. What is the percentage by mass of each element in calcium phosphate ($Ca_3(PO_4)_2$)?
12. What is the percentage by mass of each element in aluminium sulfate ($Al_2(SO_4)_3$)?
13. Hydrated copper (II) sulfate is a blue crystalline material, with formula $CuSO_4.5H_2O$. What is the percentage by mass of each element present?

Hydrated copper (II) sulfate:

14. How many moles of hydrated copper (II) sulfate ($CuSO_4.5H_2O$) are present in 25 g of the crystals?
15. When hydrated copper (II) sulfate is heated, the water of crystallization is released as steam (H_2O), leaving white crystals of anhydrous copper (II) sulfate ($CuSO_4$), according to the following reaction: $CuSO_4.5H_2O(s) \rightarrow CuSO_4(s) + 5H_2O(g)$. If 25 g of hydrated copper (II) sulfate is heated, how many moles of anhydrous copper (II) sulfate and how many moles of steam are formed? What are the masses of anhydrous copper (II) sulfate and steam?
16. A student has 20.0 g of anhydrous copper (II) sulfate crystals ($CuSO_4$) and decides to turn them blue by adding 5.0 g of water (H_2O). The student expects the following reaction to take place: $CuSO_4(s) + 5H_2O(l) \rightarrow CuSO_4.5H_2O(s)$. Has the student added enough water to react with all the anhydrous copper (II) sulfate? Which reagent is in excess and which is the limiting reactant?
17. The thermite reaction is an extremely **exothermic** (releases a lot of heat) between aluminium and iron (III) oxide: $2Al(s) + 3Fe_2O_3(s) \rightarrow 2Fe(s) + Al_2O_3(s)$. What mass of aluminium will react with 50 g of iron (III) oxide?
18. A chemist analyzes 25 g of an unidentified white crystal. He finds it contains 14.1 g of potassium, 2.2 g of carbon, and 8.7 g of oxygen. What is the empirical formula of the compound?
19. A chemist analyzes an unknown organic compound. He finds that the compound is 62.1% carbon by mass, 13.8% hydrogen by mass, and 24.1% nitrogen by mass. What is the empirical formula of the compound? If a mass spectrometer gives the m/Q of the molecular ion as 116, what must the molecular formula be?
20. Ethanol can be produced by the fermentation (using yeast or zymase) of glucose according to the following reaction: $C_6H_{12}O_6 \rightarrow 2CH_3CH_2OH + 2CO_2$. If we wish to produce 100 g of ethanol, how much glucose would be required?

Answers to Chapter 2 Questions

1. Solution:
   - $M_r = (2 \times 1.0) + (1 \times 16.0) = 18$.
   - Moles = $500 \div 18.0 = 27.8$ mol (3 significant figures).

2. Solution:
   - $M_r = (2 \times 55.8) + (3 \times 16.0) = 159.6$.
   - Moles = $500 \div 159.6 = 3.13$ mol (3 significant figures).

3. There are 2 moles of iron atoms in every mole of iron (III) oxide, therefore moles = $2 \times 3.13 = 6.27$ mol (3 significant figures).

4. There are 3 moles of oxygen atoms in every mole of iron (III) oxide, therefore moles = $3 \times 3.13 = 9.40$ mol (3 significant figures).

5. Solution:
   - $M_r = (1 \times 107.9) + (1 \times 35.5) = 143.4$.
   - Mass = $0.02 \times 143.4 = 2.87$ g (3 significant figures).

6. Solution:
   - $M_r = (1 \times 39.1) + (1 \times 54.9) + (4 \times 16.0) = 158.0$.
   - Mass = $1.87 \times 158.0 = 295$ g (3 significant figures).

7. Solution:
   - M_r Na_2CO_3 = $(2 \times 23.0) + (1 \times 12.0) + (3 \times 16.0) = 106.0$.
   - % Sodium (Na) = $(2 \times 23.0) \div 106.0 = 0.434$ then $0.434 \times 100 = 43.4\%$ (3 significant figures).
   - % Carbon (C) = $(1 \times 12.0) \div 106.0 = 0.113$ then $0.113 \times 100 = 11.3\%$ (3 significant figures).
   - % Oxygen (O) = $(3 \times 16.0) \div 106.0 = 0.453$ then $0.453 \times 100 = 45.3\%$ (3 significant figures).

8. Solution:
   - M_r AgCl = $(1 \times 107.9) + (1 \times 35.5) = 143.4$.
   - % Silver (Ag) = $(1 \times 107.9) \div 143.4 = 0.752$ then $0.752 \times 100 = 75.2\%$ (3 significant figures).
   - % Chlorine (Cl) = $(1 \times 35.5) \div 143.4 = 0.248$ then $0.248 \times 100 = 24.8\%$ (3 significant figures).

9. Solution:
   - M_r Ca_3N_2 = $(3 \times 40.1) + (2 \times 14.0) = 148.3$.
   - % Calcium (Ca) = $(3 \times 40.1) \div 148.3 = 0.811$ then $0.811 \times 100 = 81.1\%$ (3 significant figures).
   - % Nitrogen (N) = $(2 \times 14.0) \div 148.3 = 0.189$ then $0.189 \times 100 = 18.9\%$ (3 significant figures).

10. Solution:
    - M_r NH_4NO_3 = $(2 \times 14.0) + (4 \times 1.0) + (3 \times 16.0) = 80.0$.
    - % Nitrogen (N) = $(2 \times 14.0) \div 80.0 = 0.35$ then $0.35 \times 100 = 35.0\%$ (3 significant figures).
    - % Hydrogen (H) = $(4 \times 1.0) \div 80.0 = 0.05$ then $0.05 \times 100 = 5.00\%$ (3 significant figures).
    - % Oxygen (O) = $(3 \times 16.0) \div 80.0 = 0.6$ then $0.6 \times 100 = 60.0\%$ (3 significant figures).

11. Solution:
    - M_r $Ca_3(PO_4)_2$ = $(3 \times 40.1) + (2 \times 31.0) + (8 \times 16.0) = 310.3$.
    - % Calcium (Ca) = $(3 \times 40.1) \div 310.3 = 0.388$ then $0.388 \times 100 = 38.8\%$ (3 significant figures).
    - % Phosphorous (P) = $(2 \times 31.0) \div 310.3 = 0.200$ then $0.200 \times 100 = 20.0\%$ (3 significant figures).

- % Oxygen (O) = (8 × 16.0) ÷ 310.3 = 0.413 then 0.413 × 100 = 41.3% (3 significant figures).

12. Solution:
    - M_r $Al_2(SO_4)_3$ = (2 × 27.0) + (3 × 32.1) + (12 × 16.0) = 342.3.
    - % Aluminium (Al) = (2 × 27.0) ÷ 342.3 = 0.158 then 0.158 × 100 = 15.8% (3 significant figures).
    - % Sulfur (S) = (3 × 32.1) ÷ 342.3 = 0.281 then 0.281 × 100 = 28.1% (3 significant figures).
    - % Oxygen (O) = (12 × 16.0) ÷ 342.3 = 0.561 then 0.561 × 100 = 56.1% (3 significant figures).

13. Solution:
    - M_r $CuSO_4.5H_2O$ = (1 × 63.5) + (1 × 32.1) + (9 × 16.0) + (10 × 1.0) = 249.6.
    - % Copper (Cu) = (1 × 63.5) ÷ 249.6 = 0.254 then 0.254 × 100 = 25.4% (3 significant figures).
    - % Sulfur (S) = (1 × 32.1) ÷ 249.6 = 0.129 then 0.129 × 100 = 12.9% (3 significant figures).
    - % Oxygen (O) = (9 × 16.0) ÷ 249.6 = 0.577 then 0.577 × 100 = 57.7% (3 significant figures).
    - % Hydrogen (H) = (10 × 1.0) ÷ 249.6 = 0.0401 then 0.0401 × 100 = 4.01% (3 significant figures).

14. Solution:
    - M_r $CuSO_4.5H_2O$ = (1 × 63.5) + (1 × 32.1) + (9 × 16.0) + (10 × 1.0) = 249.6.
    - Moles $CuSO_4.5H_2O$ = 25 ÷ 249.6 = 0.100 mol (3 significant figures).

15. Solution:
    - The chemical equation indicates a ratio of $CuSO_4.5H_2O$: $CuSO_4$: H_2O is 1 : 1 : 5.
    - Moles of anhydrous copper (II) sulfate ($CuSO_4$) = Moles of hydrated copper (II) sulfate = 0.100 = 0.100 mol.
    - Moles of steam (H_2O) = 5 × Moles of hydrated copper (II) sulfate = 5 × 0.100 = 0.500 mol.
    - M_r anhydrous copper (II) sulfate ($CuSO_4$) = (1 × 63.5) + (1 × 32.1) + (4 × 16.0) = 159.6.
    - M_r of steam (H_2O) = (2 × 1.0) + (1 × 16.0) = 18.0.
    - Mass of anhydrous copper (II) sulfate = 0.100 × 159.6 = 16.0 g (3 significant figures).
    - Mass of steam = 0.500 × 18.0 = 9.01 g (3 significant figures).

16. Solution:
    - The chemical equation indicates a ratio of $CuSO_4$: H_2O : $CuSO_4.5H_2O$: is 1 : 5 : 1.
    - M_r of anhydrous copper (II) sulfate ($CuSO_4$) = (1 × 63.5) + (1 × 32.1) + (4 × 16.0) = 159.6.
    - Moles anhydrous copper (II) sulfate = 20.0 ÷ 159.6 = 0.125 mol.
    - M_r of water (H_2O) = (2 × 1.0) + (1 × 16.0) = 18.0.
    - Moles of water = 5.0 ÷ 18.0 = 0.278 mol.
    - Since the ratio between $CuSO_4$: H_2O is 1 : 5, this means 0.125 moles of anhydrous copper (II) sulfate could react with 5 × 0.125 = 0.625 moles of water, but there is **not** enough water available! Likewise, 0.278 moles of water could react with 0.278 ÷ 5 = 0.0556 moles of anhydrous copper (II) sulfate, but there is too much anhydrous copper (II) sulfate. Therefore, water is the limiting reactant, and anhydrous copper (II) sulfate is in excess.

17. Solution:
    - M_r Fe_2O_3 = (2 × 55.8) + (3 × 16.0) = 159.6.
    - Moles Fe_2O_3 = 50 ÷ 159.6 = 0.313 mol (3 significant figures).

- The chemical equation indicates that the ratio of Al : Fe_2O_3 is 2 : 3, therefore moles of Al = 0.313 × (2 ÷ 3) = 0.209 mol (3 significant figures).
- M_r Al = 27.0.
- Mass Al = 0.209 × 27.0 = 5.64 g (3 significant figures).

18. The empirical formula is K_2CO_3 as calculated below:

	Potassium	Carbon	Oxygen
Mass (g)	14.1	2.2	8.7
A_r	39.1	12.0	16.0
Step 1: Moles = Mass ÷ A_r	14.1 ÷ 39.1 = 0.361	2.2 ÷ 12.0 = 0.183	8.7 ÷ 16.0 = 0.544
Step 2: Ratio	0.361 ÷ 0.183 = 1.97	0.183 ÷ 0.183 = 1	0.544 ÷ 0.183 = 2.97
Step 3: Whole number ratio	2	1	3

19. The empirical formula is C_3H_8N as shown calculated below. Additionally, since we know the M_r is 116, the molecular formula must be double the empirical formula, namely $C_6H_{16}N_2$:

	Carbon	Hydrogen	Nitrogen
% of Total Mass	62.1	13.8	24.1
Mass (g) if 100g	62.1	13.8	24.1
A_r	12.0	1.0	14.0
Step 1: Moles = Mass ÷ A_r	62.1 ÷ 12.0 = 5.18	13.8 ÷ 1.0 = 13.8	24 ÷ 14.0 = 1.72
Step 2: Ratio	5.18 ÷ 1.72 = 3.01	13.8 ÷ 1.72 = 8.02	1.72 ÷ 1.72 = 1
Step 3: Whole number ratio	3	8	1

20. Solution:
- M_r of ethanol = (2 × 12.0) + (6 × 1.0) + (1 × 16.0) = 46.0.
- Moles of ethanol = 100 ÷ 46.0 = 2.17 mol (3 significant figures).
- The chemical equation indicates that the ratio of $C_6H_{12}O_6$: CH_3CH_2OH is 1 : 2, therefore moles of $C_6H_{12}O_6$ = 2.17 ÷ 2 = 1.09 mol (3 significant figures).
- M_r of $C_6H_{12}O_6$ = (6 × 12.0) + (12 × 1.0) + (6 × 16.0) = 180.0.
- Mass of $C_6H_{12}O_6$ = 1.09 × 180.0 = 196 g (3 significant figures).

Chapter 3: Solution Calculations

Very often in chemistry, we work with solutions, that is a substance dissolved in a liquid – the liquid usually being water (which would be an aqueous solution), but occasionally another solvent. It is therefore often necessary to perform calculations to find on solutions.

Volume and Concentration

When considering a solution, the two most important variables are:

(1) **Volume** – The amount of liquid measured by the space that it occupies.

Volume can be measured in m^3, dm^3 or cm^3. In chemistry, the most common units for volume are dm^3 (pronounced "decimeter cubed"). 1 dm^3 = 1000 cm^3 which corresponds to the metric unit of 1 litre (usually spelled liter in the American English).

Hence:
- Volume (dm^3) = Volume (cm^3) ÷ 1000
- Volume (cm^3) = Volume (dm^3) × 1000
- Volume (dm^3) = Volume (m^3) × 1000
- Volume (m^3) = Volume (dm^3) ÷ 1000

(2) **Concentration** – The amount of a substance (measured in moles or grams) dissolved per unit volume of the liquid.
- If measuring the amount of the dissolved substance in moles, then the concentration would usually be measured in $mol\ dm^{-3}$ (also written as mol/dm^3). The concentration of a solution in $mol\ dm^{-3}$ is sometimes known as its **Molarity** (often abbreviated to **M**), hence a 2 M solution has a concentration of 2 $mol\ dm^{-3}$.
- If measuring the amount of the dissolved substance in grams, then the concentration would usually be measured in $g\ dm^{-3}$ (also written as g/dm^3).

Solutions Formulae

We are now ready to present the solutions formulae. The two basic formulae are:
- Moles of a dissolved substance = Concentration ($mol\ dm^{-3}$) × Volume (dm^3)
- Mass of a dissolved substance (g) = Concentration ($g\ dm^{-3}$) × Volume (dm^3)

Let's try some examples:

Example: 0.15 moles of sodium chloride (NaCl) are dissolved 330 cm^3 of water. What is the concentration?

We use the formula Moles = Concentration (mol dm^{-3}) × Volume (dm^3) but rearrange it to make Concentration into the subject:
- Concentration (mol dm^{-3}) = Moles ÷ Volume (dm^3)

Since 330 cm^3 = 0.330 dm^3, we get:
- Concentration (mol dm^{-3}) = 0.15 ÷ 0.330 = 0.455 mol dm^{-3} (3 significant figures).

Example: What mass of calcium chloride (CaCl$_2$) is contained within a 2.5 dm^3 of a solution of concentration 20 g dm^{-3}?

We use the formula Mass of a dissolved substance (g) = Concentration (g dm^{-3}) × Volume (dm^3):
- Mass (g) = 20 × 2.5 = 50 g.

Converting Between Concentration Units

Since Moles = Mass (g) ÷ M_r, it follows that
- Concentration (mol dm^{-3}) = Concentration (g dm^{-3}) ÷ M_r
- Concentration (g dm^{-3}) = Concentration (mol dm^{-3}) × M_r

We can combine these with the basic formulae to produce the following formulae:
- Mass of a dissolved substance (g) = Concentration (mol dm^{-3}) × Volume (dm^3) × M_r
- Moles of a dissolved substance = Concentration (g dm^{-3}) × Volume (dm^3) ÷ M_r

Let's try applying these formulae:

Example: What is the concentration in mol dm^{-3} of a 20 g dm^{-3} solution of calcium chloride (CaCl$_2$)?

We use the formula Concentration (mol dm^{-3}) = Concentration (g dm^{-3}) ÷ M_r
- M_r of CaCl$_2$ = (1 × 40.1) + (2 × 35.5) = 111.1.
- Concentration (g dm^{-3}) = 20 ÷ 111.1 = 0.180 mol dm^{-3} (3 significant figures).

Example: What is the concentration in g dm^{-3} of a 0.455 mol dm^{-3} solution of sodium chloride (NaCl)?

We use the formula Concentration (g dm^{-3}) = Concentration (mol dm^{-3}) × M_r
- M_r of NaCl = (1 × 23.0) + (1 × 35.5) = 58.5.
- Concentration (g dm^{-3}) = 0.455 × 58.5 = 26.6 g dm^{-3} (3 significant figures).

Example: How many moles of calcium chloride (CaCl$_2$) are contained in 450 cm^3 of a 20 g dm^{-3} solution?

We use the formula Moles = Concentration (g dm^{-3}) × Volume (dm^3) ÷ M$_r$, remembering that 450 cm^3 = 0.45 dm^3:
- M$_r$ of CaCl$_2$ = (1 × 40.1) + (2 × 35.5) = 111.1.
- Moles = 20 × 0.45 ÷ 111.1 = 0.0810 mol (3 significant figures).

Example: How many grams of sodium chloride (NaCl) are contained in 9200 cm^3 of a 0.455 mol dm^{-3} solution?

We use the formula Mass = Concentration (mol dm^{-3}) × Volume (dm^3) × M$_r$, remembering that 9200 cm^3 = 9.2 dm^3:
- M$_r$ of NaCl = (1 × 23.0) + (1 × 35.5) = 58.5.
- Mass = 0.455 × 9.2 × 58.5 = 245 g (3 significant figures).

Titrations

A common type of calculation involving solutions is mixing two solutions containing different chemicals such that both dissolved chemicals entirely react with neither being in excess. This is achieved by a procedure called titration (also known as titrimetry or volumetric analysis) which involves the slow addition of one solution (known as the titrant or titrator) to another solution (known as the as the analyte or titrand) until an endpoint is reached, which is indicated by the color change of an indicator.

The most common reason for performing titrations is to determine the concentration of the analyte solution. In such cases:
- A precisely measured volume of analyte is used.
- A titrant of known concentration is used.
- The volume of titrant required to complete the reaction (called the titration volume) is precisely measured.
- The concentration of analyte can then be calculated from this information.

Titration Calculations using Moles Ratios

One way of performing titration calculations is simply to use the ratio of moles between the two reactants:

Example: We are using 25.0 cm^3 of sodium hydroxide (NaOH) of unknown concentration as the analyte. We wish to determine the analyte's concentration, so we titrate with 1.5 mol dm^{-3} hydrochloric acid (HCl). After repeated titrations, the average of concordant results (ignoring anomalies) is a titration volume of 32.4 cm^3 of hydrochloric acid. What is the concentration of the analyte?

The reaction taking place is:

HCl(aq) + NaOH(aq) → NaCl(aq) + H$_2$O(l)

Putting this together:
- 32.4 cm^3 = 0.0324 dm^3.
- 25 cm^3 = 0.025 dm^3.
- Moles HCl = Concentration (mol dm^{-3}) × Volume (dm^3) = 1.5 × 0.0324 = 0.0486 mol.
- The ratio of HCl : NaOH is 1:1 therefore Moles NaOH = 0.0486 mol.
- Moles NaOH = Concentration (mol dm^{-3}) × Volume (dm^3)
- Rearranging: Concentration (mol dm^{-3}) = Moles NaOH ÷ Volume (dm^3) = 0.0486 ÷ 0.025 = 1.94 mol dm^{-3} (3 significant figures).

Example: We are using 25.0 cm^3 of sodium hydroxide (NaOH) of unknown concentration as the analyte. We wish to determine the analyte's concentration, so we titrate with 0.9 mol dm^{-3} sulfuric acid (H$_2$SO$_4$). After repeated titrations, the average of concordant results (ignoring anomalies) is a titration volume of 17.1 cm^3 of sulfuric acid. What is the concentration of the analyte?

The reaction taking place is:

H$_2$SO$_4$(aq) + 2NaOH(aq) → Na$_2$SO$_4$(aq) + 2H$_2$O(l)

Putting this together:
- 17.1 cm^3 = 0.0171 dm^3.
- 25 cm^3 = 0.025 dm^3.
- Moles H$_2$SO$_4$ = Concentration (mol dm^{-3}) × Volume (dm^3) = 0.9 × 0.0171 = 0.0154 mol.
- The ratio of H$_2$SO$_4$: NaOH is 1:2 therefore Moles NaOH = 0.0154 × 2 = 0.0308 mol.
- Moles NaOH = Concentration (mol dm^{-3}) × Volume (dm^3)
- Rearranging: Concentration (mol dm^{-3}) = Moles NaOH ÷ Volume (dm^3) = 0.0308 ÷ 0.025 = 1.23 mol dm^{-3} (3 significant figures).

Titration Calculations in One Step

While there is nothing wrong with performing the titration calculations as described above using the moles ratios between the different reactants, it is possible to include the ratio of moles directly within the formulae used in calculations. This can be a good time saver if you have many calculations to perform.

In general, if we have reactants 1 and 2, for all titrations:
- Concentration1 (mol dm^{-3}) × Volume1 (dm^3) ÷ n1 = Concentration2 (mol dm^{-3}) × Volume2 (dm^3) ÷ n2

Where n1 and n2 are the numbers that appear before a reactant in the chemical equation (if no number is present, treat this as an implicit 1).

Let's repeat the examples we just did above, but using this method:

Example: We are using 25.0 cm³ of sodium hydroxide (NaOH) of unknown concentration as the analyte. We wish to determine the analyte's concentration, so we titrate with 1.5 mol dm^{-3} hydrochloric acid (HCl). After repeated titrations, the average of concordant results (ignoring anomalies) is a titration volume of 32.4 cm³ of hydrochloric acid. What is the concentration of the analyte?

The reaction taking place is:

HCl(aq) + NaOH(aq) → NaCl(aq) + H$_2$O(l)

So:
- Reactant1 = HCl.
- Concentration1 = 1.5 mol dm^{-3}.
- Volume1 = 32.4 cm³ = 0.0324 dm³.
- n1 = 1.
- Reactant2 = NaOH.
- Concentration2 = unknown.
- Volume2 = 25.0 cm³ = 0.025 dm³.
- n2 = 1

Rearranging the formula, we get:
- Concentration2 (mol dm^{-3}) = Concentration1 (mol dm^{-3}) × Volume1 (dm³) × n2 ÷ (n1 × Volume2 (dm³))

Hence:
- Concentration NaOH = 1.5 × 0.0324 × 1 ÷ (1 × 0.025) = 1.94 mol dm^{-3} (3 significant figures).

Example: We are using 25.0 cm³ of sodium hydroxide (NaOH) of unknown concentration as the analyte. We wish to determine the analyte's concentration, so we titrate with 0.9 mol dm^{-3} sulfuric acid (H$_2$SO$_4$). After repeated titrations, the average of concordant results (ignoring anomalies) is a titration volume of 17.1 cm³ of sulfuric acid. What is the concentration of the analyte?

The reaction taking place is:

H$_2$SO$_4$(aq) + 2NaOH(aq) → Na$_2$SO$_4$ (aq) + 2H$_2$O(l)

So:
- Reactant1 = H_2SO_4.
- Concentration1 = 0.9 mol dm^{-3}.
- Volume1 = 17.1 cm^3 = 0.0171 dm^3.
- n1 = 1.
- Reactant2 = NaOH.
- Concentration2 = unknown.
- Volume2 = 25.0 cm^3 = 0.025 dm^3.
- n2 = 2

Rearranging the formula, we get:
- Concentration2 (mol dm-3) = Concentration1 (mol dm-3) × Volume1 (dm3) × n2 ÷ (n1 × Volume2 (dm^3))

Hence:
- Concentration NaOH = 0.9 × 0.0171 × 2 ÷ (1 × 0.025) = 1.23 mol dm^{-3} (3 significant figures).

Questions

1. 2 g of potassium chloride (KCl) is dissolved in 250 cm^3 of water. What is the concentration in g dm^{-3}? What is the concentration in mol dm^{-3}?

2. 33 g of sodium chloride (NaCl) is dissolved in 1350 cm^3 of water. What is the concentration in g dm^{-3}? What is the concentration in mol dm^{-3}?

3. How many moles of copper (II) sulfate ($CuSO_4$) are present in 350 cm^3 of a solution with concentration of 0.2 mol dm^{-3}? What mass of copper is present?

4. A chemist wishes to make 250 cm^3 of a 1.0 mol dm^{-3} solution of potassium chloride (KCl). What mass of potassium chloride should be mixed with the water?

5. A chemist wishes to make 500 cm^3 of a 1.0 mol dm^{-3} solution of sodium chloride (NaCl). What mass of sodium chloride should be mixed with the water?

6. A chemist wishes to dissolve 5 g of calcium chloride ($CaCl_2$) to make a solution of concentration 1 g dm^{-3}. What volume of water should be used?

7. A chemist wishes to dissolve 5 g of calcium chloride ($CaCl_2$) to make a solution of concentration 0.1 mol dm^{-3}. What volume of water should be used?

8. What mass of potassium permanganate ($KMnO_4$) needs to be dissolved in each dm^3 of water to make a solution with concentration 0.3 mol dm^{-3}?

9. What mass of potassium nitrate (KNO_3) needs to be dissolved in each dm^3 of water to make a solution with concentration 0.2 mol dm^{-3}?

10. What mass of calcium nitrate ($Ca(NO_3)_2$) needs to be dissolved in each dm^3 of water to make a solution with concentration 2.0 mol dm^{-3}?

11. 25 g of potassium nitrate (KNO_3) is dissolved in 650 cm^3 of water. What is the concentration in g dm^{-3}? What is the concentration in mol dm^{-3}?

12. Which contains more dissolved sodium chloride (NaCl): 120 cm³ of a 7 g dm⁻³ solution, or 140 cm³ of a 0.1 mol dm⁻³ solution?

13. How many moles of sodium carbonate (Na_2CO_3) are there in 325 cm³ of 0.15 mol dm⁻³ solution?

14. What mass of calcium chloride ($CaCl_2$) would need to be dissolved in 100 cm³ of water to make 0.20 mol dm⁻³ solution?

15. 32 g of potassium permanganate ($KMnO_4$) are to be dissolved in water. If we wish to create a solution with a concentration of 0.1 mol dm⁻³, what volume of water should be used?

Potassium permanganate solution:

16. At a temperature of 20°C, the maximum solubility of potassium permanganate ($KMnO_4$) in water is 64.0 g dm⁻³. What is this in mol dm⁻³?

17. 25 cm³ of a solution of unknown concentration of potassium hydroxide (KOH) titrates with 43.2 cm³ of 0.4 mol dm⁻³ nitric acid (HNO_3). The reaction that takes place is KOH(aq) + HNO_3(aq) → KNO_3(aq) + H_2O (l). What is the concentration of the potassium hydroxide solution?

18. 25 cm³ of a solution of unknown concentration of sodium hydroxide (NaOH) titrates with 17.8 cm³ of 0.8 mol dm⁻³ sulfuric acid (H_2SO_4). The reaction that takes place is 2NaOH(aq) + H_2SO_4(aq) → Na_2SO_4 (aq) + 2H_2O (l). What is the concentration of the sodium hydroxide solution?

19. A chemist wishes to react 32.0 cm³ of a 0.85 mol dm⁻³ solution of sodium hydroxide (NaOH) with a 2.00 mol dm⁻³ solution of hydrochloric acid (HCl). The reaction that takes place is NaOH(aq) + HCl(aq) → NaCl(aq) + H_2O(l). What volume of hydrochloric acid is needed if neither reactant is to be in excess?

20. A chemist wishes to react 48.0 cm³ of a 1.40 mol dm⁻³ solution of sodium hydroxide (NaOH) with a 2.00 mol dm⁻³ solution of sulfuric acid (H_2SO_4). The reaction that takes place is 2NaOH(aq) + H_2SO_4(aq) → Na_2SO_4(aq) + 2H_2O(l). What volume of sulfuric acid is needed if neither reactant is to be in excess?

Answers to Chapter 3 Questions

1. Solution:
   - M_r KCl = (1 × 39.1) + (1 × 35.5) = 74.6.
   - 250 cm^3 = 0.250 dm^3.
   - Concentration (g dm^{-3}) = Mass ÷ Volume = 2 ÷ 0.250 = 8.00 g dm^{-3} (3 significant figures).
   - Concentration (mol dm^{-3}) = Mass ÷ (Volume × M_r) = 2 ÷ (0.250 × 74.6) = 0.107 mol dm^{-3} (3 significant figures).

2. Solution:
   - M_r NaCl = (1 × 23.0) + (1 × 35.5) = 58.8.
   - 1350 cm^3 = 1.350 dm^3.
   - Concentration (g dm^{-3}) = Mass ÷ Volume = 33 ÷ 1.350 = 24.4 g dm^{-3} (3 significant figures).
   - Concentration (mol dm^{-3}) = Mass ÷ (Volume × M_r) = 33 ÷ (1.350 × 58.5) = 0.418 mol dm^{-3} (3 significant figures).

3. Solution:
   - A_r Cu = 63.5.
   - 350 cm^3 = 0.350 dm^3.
   - Moles CuSO$_4$ = Concentration (mol dm^{-3}) × Volume (dm^3) = 0.2 × 0.350 = 0.0700 mol (3 significant figures).
   - Since each CuSO$_4$ contains 1 Cu atom, Moles Cu = 0.0700 mol (3 significant figures).
   - Mass Cu = Moles × A_r = 0.0700 × 63.5 = 4.45 g (3 significant figures).

4. Solution:
   - M_r KCl = (1 × 39.1) + (1 × 35.5) = 74.6.
   - 250 cm^3 = 0.250 dm^3.
   - Moles = Concentration (mol dm^{-3}) × Volume (dm^3) = 1.0 × 0.250 = 0.250 mol (3 significant figures).
   - Mass KCl = Moles × M_r = 0.250 × 74.6 = 18.7 g (3 significant figures).

5. Solution:
   - M_r NaCl = (1 × 23.0) + (1 × 35.5) = 58.5.
   - 500 cm^3 = 0.500 dm^3.
   - Moles = Concentration (mol dm^{-3}) × Volume (dm^3) = 1.0 × 0.500 = 0.500 mol (3 significant figures).
   - Mass NaCl = Moles × M_r = 0.500 × 58.5 = 29.3 g (3 significant figures).

6. Solution, by rearranging Concentration (g dm^{-3}) = Mass ÷ Volume:
   - Volume = Concentration (g dm^{-3}) × Mass = 1 × 5 = 5 dm^3.

7. Solution:
   - M_r CaCl$_2$ = (1 × 40.1) + (2 × 35.5) = 111.1.
   - Moles CaCl$_2$ = Mass ÷ M_r = 5 ÷ 111.1 = 0.0450 mol (3 significant figures)
   - Volume = Moles ÷ Concentration (mol dm^{-3}) = 0.0450 ÷ 0.1 = 0.450 dm^3 = 450 cm^3 (3 significant figures).

8. Solution:
   - M_r KMnO$_4$ = (1 × 39.1) + (1 × 54.9) + (4 × 16.0) = 158.0.

- Volume = 1 dm^3.
- Moles KMnO$_4$ = Concentration (mol dm^{-3}) × Volume (dm^3) = 0.3 × 1 = 0.3 mol.
- Mass KMnO$_4$ = Moles × M$_r$ = 0.3 × 158.0 = 47.4 g (3 significant figures).

9. Solution:
   - M$_r$ KNO$_3$ = (1 × 39.1) + (1 × 14.0) + (3 × 16.0) = 101.1.
   - Volume = 1 dm^3.
   - Moles KNO$_3$ = Concentration (mol dm^{-3}) × Volume (dm^3) = 0.2 × 1 = 0.2 mol.
   - Mass KNO$_3$ = Moles × M$_r$ = 0.2 × 101.1 = 20.2 g (3 significant figures).

10. Solution:
    - M$_r$ Ca(NO$_3$)$_2$ = (1 × 40.1) + (2 × 14.0) + (6 × 16.0) = 164.1.
    - Volume = 1 dm^3.
    - Moles Ca(NO$_3$)$_2$ = Concentration (mol dm^{-3}) × Volume (dm^3) = 2.0 × 1 = 2.0 mol.
    - Mass Ca(NO$_3$)$_2$ = Moles × M$_r$ = 2.0 × 164.1 = 328 g (3 significant figures).

11. Solution:
    - M$_r$ KNO$_3$ = (1 × 39.1) + (1 × 14.0) + (3 × 16.0) = 101.1.
    - 650 cm^3 = 0.650 dm^3.
    - Concentration (g dm^{-3}) = Mass ÷ Volume = 25 ÷ 0.650 = 38.5 g dm^{-3} (3 significant figures).
    - Concentration (mol dm^{-3}) = Mass ÷ (Volume × M$_r$) = 25 ÷ (0.650 × 101.1) = 0.380 mol dm^{-3} (3 significant figures).

12. Solution:
    - M$_r$ NaCl = (1 × 23.0) + (1 × 35.5) = 58.5.
    - 120 cm^3 = 0.120 dm^3.
    - 140 cm^3 = 0.140 dm^3.
    - Mass NaCl in 120 cm^3 solution = Concentration (g dm^{-3}) × Volume (dm^3) = 7 × 0.120 = 0.84 g.
    - Mass NaCl in 140 cm^3 solution = Concentration (mol dm^{-3}) × Volume (dm^3) × M$_r$ = 0.1 × 0.140 × 58.5 = 0.819 g.
    - Therefore the 120 cm^3 solution contains slightly more dissolved sodium chloride.

13. Solution:
    - 325 cm^3 = 0.325 dm^3.
    - Moles = Concentration (mol dm^{-3}) × Volume (dm^3) = 0.15 × 0.325 = 0.0488 mol (3 significant figures).

14. Solution:
    - M$_r$ CaCl$_2$ = (1 × 40.1) + (2 × 35.5) = 111.1.
    - 100 cm^3 = 0.100 dm^3.
    - Mass = Concentration (mol dm^{-3}) × Volume (dm^3) × M$_r$ = 0.20 × 0.100 × 111.1 = 2.22 g (3 significant figures).

15. Solution:
    - M$_r$ KMnO$_4$ = (1 × 39.1) + (1 × 54.9) + (4 × 16.0) = 158.0.
    - Volume = Mass ÷ (Concentration (mol dm^{-3}) × M$_r$) = 32 ÷ (0.1 × 158.0) = 2.03 dm^3 = 2030 cm^3 (3 significant figures).

16. Solution:

- M_r KMnO$_4$ = (1 × 39.1) + (1 × 54.9) + (4 × 16.0) = 158.0.
- Concentration (mol dm^{-3}) = Concentration (g dm^{-3}) ÷ M_r = 64.0 ÷ 158.0 = 0.405 mol dm^{-3}.

17. Solution (using the rearranged single step formula):
    - Reactant1 = HNO$_3$.
    - Concentration1 = 0.4 mol dm^{-3}.
    - Volume1 = 43.2 cm^3 = 0.0432 dm^3.
    - n1 = 1.
    - Reactant2 = KOH.
    - Concentration2 = unknown.
    - Volume2 = 25 cm^3 = 0.025 dm^3.
    - n2 = 1.
    - Concentration2 (mol dm^{-3}) = Concentration1 (mol dm^{-3}) × Volume1 (dm^3) × n2 ÷ (n1 × Volume2 (dm^3)) = 0.4 × 0.0432 × 1 ÷ (1 × 0.025) = 0.691 mol dm^{-3}.

18. Solution (using the rearranged single step formula):
    - Reactant1 = H$_2$SO$_4$.
    - Concentration1 = 0.8 mol dm^{-3}.
    - Volume1 = 17.8 cm^3 = 0.0178 dm^3.
    - n1 = 1.
    - Reactant2 = NaOH.
    - Concentration2 = unknown.
    - Volume2 = 25 cm^3 = 0.025 dm^3.
    - n2 = 2.
    - Concentration2 (mol dm^{-3}) = Concentration1 (mol dm^{-3}) × Volume1 (dm^3) × n2 ÷ (n1 × Volume2 (dm^3)) = 0.8 × 0.0178 × 2 ÷ (1 × 0.025) = 1.14 mol dm^{-3}.

19. Solution (using the rearranged single step formula):
    - Reactant1 = NaOH.
    - Concentration1 = 0.85 mol dm^{-3}.
    - Volume1 = 32.0 cm^3 = 0.0320 dm^3.
    - n1 = 1.
    - Reactant2 = HCl.
    - Concentration2 = 2.00 mol dm^{-3}.
    - Volume2 = unknown.
    - n2 = 1.
    - Volume2 (dm^3) = Concentration1 (mol dm^{-3}) × Volume1 (dm^3) × n2 ÷ (n1 × Concentration2 (mol dm^{-3})) = 0.85 × 0.0320 × 1 ÷ (1 × 2.00) = 0.0136 dm^3 = 13.6 cm^3 (3 significant figures).

20. Solution (using the rearranged single step formula):
    - Reactant1 = NaOH.
    - Concentration1 = 1.40 mol dm^{-3}.
    - Volume1 = 48.0 cm^3 = 0.0480 dm^3.
    - n1 = 2.

- Reactant2 = H_2SO_4.
- Concentration2 = 2.00 mol dm^{-3}.
- Volume2 = unknown.
- n2 = 1.
- Volume2 (dm^3) = Concentration1 (mol dm^{-3}) × Volume1 (dm^3) × n2 ÷ (n1 × Concentration2 (mol dm^{-3})) = 1.40 × 0.0480 × 1 ÷ (2 × 2.00) = 0.0168 dm^3 = 16.8 cm^3 (3 significant figures).

Chapter 4: Gas Calculations
In this chapter, we will discuss calculations involving gases.

Ideal Gases
Gases consist of particles, widely spaced, moving with relatively high velocities. Pressure is caused by collisions of the gas particles with the walls of the container.
- If we add more gas particles within the same space, this would result in more collisions with the walls of the container and thus increase the pressure. Conversely, removing gas particles from the space would decrease the pressure
- If we keep the number of gas particles constant, and increase the volume of the container, this reduces the number of collisions with the walls of the container and decreases the pressure. Conversely, decreasing the volume of the container, increases the number of collisions with the walls of the container and increases the pressure.
- If we raise the temperature, the kinetic energy of the gas particles increases (meaning that they are moving faster), which increases both the number of collisions with the walls of the container and the force of each collision, thus increasing the pressure. Conversely, decreasing the temperature decreases the kinetic energy of the gas particles (meaning that they are moving slower), which decreases the number of collisions with the walls of the container and the force of each collision, thus decreasing the pressure.

All above factors are true for every gas. However different types of gases also differ from each other in two ways:
- The amount of space occupied by the gas particles themselves: For example, a hydrogen H_2 molecule (consisting of two hydrogen atoms), would be much smaller than a methane CH_4 molecule (consisting of four hydrogen atoms and one carbon atom).
- The interactions between the gas particles: For example, a polar molecule such as fluoromethane CH_3F will have much stronger intermolecular forces than a non-polar molecule like methane CH_4.

Because of the differences between various gases, it might seem that it would be difficult to develop a general set of numerical formulae that would apply to all gases. However, it is possible to create formulae describing an ideal gas – and for most real gases under typical conditions, the errors are quite small.

What is an ideal gas and why can it be a good approximation for real gases?
- In an ideal gas, we assume that the gas particles occupy no space at all (sometimes they are described as point particles). This is usually a very good approximation for real gases, because gas molecules are generally small and very far apart – occupying only a tiny proportion of the available space.
- In an ideal gas, we assume that there are no mathematically significant interactions between the particles (sometimes the particles are said to have perfectly elastic collisions, meaning

that no kinetic energy is lost in collisions between the particles). This is usually a very good approximation for real gases, because gas molecules generally only have weak intermolecular forces between them, and because they are generally spaced far apart, that they collide relatively infrequently.

Of course, how similar a real gas is to an ideal gas can vary. For example, in real gases at a high pressure more of the available space is occupied by gas molecules than in real gases at low pressures. Likewise, real gases at high pressures or high temperatures would experience more collisions between the gas molecules than when at a low pressure or low temperature. And so… In general, the greater the differences between the real gas and an ideal gas, the larger will be any numerical errors in calculations based on ideal gas assumptions.

Avogadro's Law
Avogadro's Law (sometimes called Avogadro's hypothesis or Avogadro's principle, and in France known as Ampère's hypothesis, the Avogadro-Ampère hypothesis or the Ampère-Avogadro hypothesis) is an experimental gas law first described by Amedeo Avogadro (whom we encountered in Chapter 1) in 1811. While **not** exactly true for any real gas, Avogadro's Law is correct for ideal gases and therefore very close to being correct for real gases under typical conditions.

Avogadro's Law states:
- Under equal conditions of pressure and temperature, equal volumes of **all** gases contain equal numbers of molecules.

This can also be expressed:
- Volume ÷ Number of Particles = Constant

Example: A given quantity of hydrogen occupies 1.5 dm³ at a certain pressure and temperature. With what volume of oxygen (O_2), at the same pressure and temperature, will it react, according to the following chemical equation $2H_2(g) + O_2(g) \rightarrow 2H_2O(g)$.

Since the ratio of $H_2 : O_2$ is 2 : 1, we need have half as many oxygen molecules as we have hydrogen molecules. Therefore, the volume of oxygen required is half that of hydrogen. Hence volume of oxygen = 1.5 ÷ 2 = 0.75 dm³.

Example: 0.009 moles of neon gas occupies a volume of 207 cm³ at a given temperature and pressure. What volume will 0.012 moles of neon occupy at the same temperature and pressure?

The number of neon molecules was increased by a factor 0.012 ÷ 0.009 = 1.33. Therefore, the volume of the gas will increase by the same factor. Final volume of neon gas = 207 × 1.33 = 276 cm³.

Example: A cylinder containing a piston of volume of 500 cm^3 contains 0.4 g of helium gas. More helium is added, the piston moves, and the volume increases to 750 cm^3. The pressure and temperature are kept constant. What mass of helium does the piston now contain?

The volume was increased by a factor of 750 ÷ 500 = 1.5. Therefore, the number of helium molecules was also increased by a factor of 1.5. Since the total mass of the helium molecules is proportional to the number of molecules, the total mass is also increased by a factor of 1.5. Therefore, final mass of helium = 0.4 × 1.5 = 0.6 g.

Ideal Gases at Room Temperature and Pressure

At **room temperature and pressure** (**r.t.p**) which is defined as a temperature of 25°C (298.15K) and a pressure of 101.325 kPa (also known as 1 atmosphere), 1 mole of an ideal gas occupies a volume of 24 dm^3, hence:
- Moles = Volume (dm^3) ÷ 24

Provided the conditions match r.t.p, we can use this formula to calculate moles from gas volumes and vice-versa.

Example: 7 g of calcium carbonate ($CaCO_3$) are heated using a Bunsen burner. The calcium carbonate undergoes thermal decomposition to produce calcium oxide (CaO) and carbon dioxide (CO_2) according to the following chemical equation:

$CaCO_3$(s) → CaO(s) + CO_2(g)

What is the volume of carbon dioxide that can be produced at r.t.p?

Solution:
- M_r of $CaCO_3$ = (1 × 40.1) + (1 × 12.0) + (3 × 16.0) = 100.1.
- Moles of $CaCO_3$ = 7 ÷ 100.1 = 0.0699 moles (3 significant figures).
- Ratio of $CaCO_3$: CO_2 is 1 : 1, hence moles CO_2 = 0.0699 (3 significant figures).
- Volume of CO_2 = 0.0699 × 24 = 1.68 dm^3 (3 significant figures).

Ideal Gases at Standard Temperature and Pressure

At **standard temperature and pressure** (**s.t.p**) which is defined as a temperature of 0°C (273.15K) and a pressure of 101.325 kPa (also known as 1 atmosphere), 1 mole of an ideal gas occupies a volume of 22.4 dm^3, hence:
- Moles = Volume (dm^3) ÷ 22.4

Provided the conditions match s.t.p, we can use this formula to calculate moles from gas volumes and vice-versa. Note: Despite having the name of "standard temperature and pressure", nowadays it is generally less common to do calculations under s.t.p conditions than under r.t.p conditions.

Example: Lithium (Li) reacts with oxygen (O_2) to produce lithium oxide (Li_2O) according to the following chemical equation:

$4Li(s) + O_2(g) \rightarrow 2Li_2O(s)$

What volume of oxygen will react with 10 g of lithium at s.t.p?

Solution:
- A_r of Li = 6.9
- Moles of Li = 10 ÷ 6.9 = 1.45 moles (3 significant figures).
- Ratio Li : O_2 is 4 : 1, hence moles O_2 = 1.45 ÷ 4 = 0.362 moles (3 significant figures).
- Volume of O_2 = 0.362 × 22.4 = 8.12 dm³ (3 significant figures).

Ideal Gas Law
While it is useful to be able to perform calculations at r.t.p or s.t.p, we also often need to perform calculations at other pressures and temperatures.

The Ideal Gas Law allows us to do so. The formula is:
- Ideal Gas Law: Moles = (Pressure (Pa) × Volume (m³)) ÷ (R (J K^{-1} mol^{-1}) × Temperature (K))

Or, omitting units:
- Moles = (Pressure × Volume) ÷ (R × Temperature)

Or, in short:
- n = PV ÷ RT

Three **Very Important Notes:**
- In the above formula, R is a constant known as the Ideal Gas Constant which has a value of 8.314 J K^{-1} mol^{-1}.
- Volumes in this formula are expressed in m³ – and **not** dm³. To convert between the two, remember that 1 m³ = 1000 dm³.
- Pressures in this formula are expressed in Pascals (symbol: Pa) – and **not** kiloPascals (symbol: kPa). To convert between the two, remember 1 kPa = 1000 Pa.
- Temperature must **always** be measured using the Kelvin scale. To convert a temperature in kelvin (symbol: K). To convert a temperature in Celsius (symbol: °C) to kelvin, simply add 273.15. To convert a temperature in kelvin to Celsius, simply subtract 273.15.

You Can Do Chemistry: Moles & Stoichiometry

Example: According to the Ideal Gas Law, what volume would 1 mole of gas occupy at a temperature of 25°C and a pressure of 101.325 kPa?

Solution:
- Rearranging the Ideal Gas Law, we get Volume = Moles × Temperature × R ÷ Pressure.
- Moles = 1 mol.
- Temperature = 25°C = 25 + 273.15 = 298.15 K.
- R = 8.314 J K^{-1} mol^{-1}.
- Pressure = 101.325 kPa = 101325 Pa.
- Therefore: Volume (m^3) = 1 × 298.15 × 8.314 ÷ 101325 = 0.024464 m^3 = 24.464 dm^3 ≈ 24.5 dm^3 (3 significant figures).

Example: A helium balloon containing 1 mole of the gas rises to a height of around 11,000 meters above sea-level. The temperature at this altitude is -60°C and the pressure is 23.8 kPa. What is the volume of the balloon at this altitude?

Solution:
- Rearranging the Ideal Gas Law, we get Volume = Moles × Temperature × R ÷ Pressure.
- Moles = 1 mol.
- Temperature = -60°C = -60 + 273.15 = 213.15 K.
- R = 8.314 J K^{-1} mol^{-1}.
- Pressure = 23.8 kPa = 23800 Pa.
- Therefore: Volume (m^3) = 1 × 213.15 × 8.314 ÷ 23800 = 0.074459 m^3 = 74.459 dm^3 ≈ 74.5 dm^3 (3 significant figures).

Example: A cylinder containing oxygen (O_2) has a volume of 460 liters, a pressure of 23000 kPa, and is at a temperature of 18°C. How many moles of oxygen are contained in the cylinder? What is the mass of the oxygen?

Solution:
- Moles = (Pressure × Volume) ÷ (R × Temperature)
- Pressure = 23000 kPa = 23000000 Pa.
- Volume = 460 liters = 460 dm^3 = 0.460 m^3.
- R = 8.314 J K^{-1} mol^{-1}.
- Temperature = 18°C = 18 + 273.15 = 291.15 K.
- Therefore Moles = (23000000 × 0.460) ÷ (8.314 × 291.15) = 4370 moles (3 significant figures).
- M_r O_2 = (2 × 16.0) = 32.0.
- Mass = Moles × M_r = 4370 × 32.0 = 139865 g = 139.865 kg = 140 kg (3 significant figures).

Questions

1. 5 dm^3 of methane (CH_4) undergoes complete combustion in oxygen (O_2) to produce carbon monoxide and water vapor according to the following chemical equation: $2CH_4(g) + 4O_2(g)$ →

You Can Do Chemistry: Moles & Stoichiometry

$2CO_2(g) + 4H_2O(g)$. What volume of oxygen is need for this reaction? Using the same assumption, what is the total volume of products? Why might this assumption **not** be valid?

2. A child's balloon of volume 1055 cm³ is filled with helium (He) at r.t.p. How many moles of helium are used? What is the mass of the helium?

3. A 10 g lump of sodium metal is dropped in a large bucket of water (Note: Do **not** try this!). The following chemical reaction takes place: $2Na(s) + 2H_2O(l) \rightarrow 2NaOH(aq) + H_2(g)$. Given that water is in excess, what volume of hydrogen will be produced at r.t.p?

4. What volume will 2.5 g of carbon dioxide (CO_2) gas occupy at r.t.p?

5. How many moles of carbon dioxide (CO_2) gas are contained in 1800 cm³ container at r.t.p? What is the mass of the carbon dioxide?

6. What volume will 0.5 g of oxygen (O_2) gas occupy at r.t.p?

7. How many moles of nitrogen (N_2) gas are contained in 4 dm³ container at s.t.p? What is the mass of the nitrogen?

8. How many moles of oxygen (O_2) gas are contained in 4 dm³ container at s.t.p? What is the mass of the oxygen?

9. How many moles of carbon dioxide (CO_2) gas are contained in 4 dm³ container at s.t.p? What is the mass of the carbon dioxide?

10. Excess calcium carbonate ($CaCO_3$) is reacted with 50 cm³ of 0.9 mol dm⁻³ hydrochloric acid. The following chemical reaction takes place: $CaCO_3(s) + 2HCl(aq) \rightarrow CaCl_2(aq) + H_2O(l) + CO_2(g)$. What volume of carbon dioxide (CO_2) is produced by the reaction if performed at r.t.p?

11. What volume will 25 g of water (H_2O) occupy if heated to 110°C on a day when the pressure is 100 kPa?

12. What volume will 7 g of oxygen (O_2) occupy at a temperature of 350K and a pressure of 200 kPa?

13. What volume will 7 g of oxygen (O_2) occupy at a temperature of -40°C and a pressure of 100 kPa?

14. What volume will 0.5 moles of hydrogen iodide (HI) occupy at a temperature of 300K and 100 kPa?

15. Assuming a pressure of 100 kPa, at what temperature will 20 g of nitrogen (N_2) gas occupy a volume of 32 dm³?

16. 10 g of calcium nitrate is heated and undergoes thermal decomposition. The chemical reaction that takes place is: $2Ca(NO_3)_2(s) \rightarrow 2CaO(s) + 4NO_2(g) + O_2(g)$. What is the volume of gas produced assuming r.t.p?

17. 1300 cm³ of hydrogen chloride (HCl) gas at temperature of 320K and 100 kPa pressure is dissolved in 5000 cm³ of water (H_2O). What is the concentration of the resulting solution of HCl?

18. A 15 liter container contains 27.5 g of an unknown gas at r.t.p. Is the gas hydrogen (H_2), helium (He), nitrogen (N_2), oxygen (O_2) or carbon dioxide (CO_2)?

19. 10 g of aluminium sulfide (Al_2S_3) is accidentally exposed to water, releasing the smelly and toxic gas, hydrogen sulfide (H_2S). The chemical equation for the reaction is $Al_2S_3(s) + 6H_2O(l) \rightarrow 2Al(OH)_3(s) + 3H_2S(g)$. What is the maximum volume of hydrogen sulfide that could be produced at r.t.p?

20. A reactor performing the Haber process (the manufacture of ammonia gas from nitrogen and hydrogen gases) has a volume of 800 m^3. The reaction is carried out at 450°C and at a pressure 21 MPa (21000 kPa). How many moles of gas does the reactor vessel contain?

Answers to Chapter 4 Questions

1. Solution:
    - According to Avogadro's Law, under equal conditions of pressure and temperature, equal volumes of gases contain equal numbers of molecules.
    - Therefore, assuming that the methane and oxygen are at the same temperature and volume, since the ratio of $CH_4 : O_2$ molecules is 2 : 3, the ratio of their volumes should also be 2 : 3. So, the volume of oxygen needed is $5 \times (3 \div 2) = 7.5$ dm^3.
    - The ratio of CH_4 : Gaseous product molecules is 2 : 2 + 4 = 2 : 6, so assuming the gaseous products are at the same pressure and temperature as the methane, their volume should be $5 \times (6 \div 2) = 15$ dm^3.
    - The assumption that the gaseous products are at the same temperature and pressure as the initial methane/oxygen mix could easily **not** be correct, because the combustion of methane would release heat (increasing the temperature).

2. Solution:
    - A_r He = 4.0.
    - Volume = 1055 cm^3 = 1.055 dm^3.
    - Moles = Volume ÷ 24 = 1.055 ÷ 24 = 0.0440 mol (3 significant figures).
    - Mass = Moles × A_r = 0.440 × 4.0 = 0.176 g (3 significant figures).

3. Solution:
    - A_r Na = 23.0.
    - Moles Na = Mass ÷ A_r = 10 ÷ 23.0 = 0.435 mol.
    - Ratio Na : H_2 is 2 : 1, therefore moles H_2 = 0.435 ÷ 2 = 0.217 mol (3 significant figures).
    - Volume H_2 = Moles × 24 = 0.217 × 24 = 5.22 dm^3 (3 significant figures).

4. Solution:
    - M_r CO_2 = (1 × 12.0) + (2 × 16.0) = 44.0.
    - Moles CO_2 = Mass ÷ M_r = 2.5 ÷ 44.0 = 0.0568 mol (3 significant figures).
    - Volume CO_2 = Moles × 24 = 0.0568 × 24 = 1.36 dm^3 (3 significant figures).

5. Solution:
    - M_r CO_2 = (1 × 12.0) + (2 × 16.0) = 44.0.
    - Volume = 1800 cm^3 = 1.800 dm^3.
    - Moles CO_2 = Volume ÷ 24 = 1.800 ÷ 24 = 0.075 mol.
    - Mass CO_2 = Moles × M_r = 0.075 × 44.0 = 3.3 g.

6. Solution:
    - M_r O_2 = (2 × 16.0) = 32.0.
    - Moles O_2 = Mass ÷ M_r = 0.5 ÷ 32.0 = 0.0156 mol (3 significant figures).
    - Volume O_2 = Moles × 24 = 0.0156 × 24 = 0.375 dm^3 (3 significant figures).

7. Solution:
    - M_r N_2 = (2 × 14.0) = 28.0.
    - Moles N_2 = Volume ÷ 22.4 = 4 ÷ 22.4 = 0.179 mol (3 significant figures).
    - Mass N_2 = Moles × M_r = 0.179 × 28.0 = 5.00 g (3 significant figures).

8. Solution:
    - $M_r\ O_2 = (2 \times 16.0) = 32.0$.
    - Moles O_2 = Volume ÷ 22.4 = 4 ÷ 22.4 = 0.179 mol (3 significant figures).
    - Mass O_2 = Moles × M_r = 0.179 × 32.0 = 5.71 g (3 significant figures).
9. Solution:
    - $M_r\ CO_2 = (1 \times 12.0) + (2 \times 16.0) = 44.0$.
    - Moles CO_2 = Volume ÷ 22.4 = 4 ÷ 22.4 = 0.179 mol (3 significant figures).
    - Mass CO_2 = Moles × M_r = 0.179 × 44.0 = 7.86 g (3 significant figures).
10. Solution:
    - Volume = 50 cm³ = 0.050 dm³.
    - Moles HCl = Concentration × Volume = 0.9 × 0.050 = 0.045 mol.
    - Ratio HCl : CO_2 is 2 :1, therefore moles CO_2 = 0.045 ÷ 2 = 0.0225 mol.
    - Volume CO_2 = Moles × 24 = 0.54 dm³.
11. Solution:
    - $M_r\ H_2O = (2 \times 1.0) + (1 \times 16.0) = 18.0$.
    - Temperature = 110°C = (110 + 273.15) = 383.15K.
    - Pressure = 100 kPa = 100000 Pa.
    - Moles H_2O = Mass ÷ M_r = 25 ÷ 18.0 = 1.39 mol (3 significant figures).
    - Rearranging the Ideal Gas Law, we get Volume = Moles × Temperature × R ÷ Pressure.
    - Volume H_2O = 1.39 × 383.15 × 8.314 ÷ 100000 = 0.0442 m³ = 44.2 dm³ (3 significant figures).
12. Solution:
    - $M_r\ O_2 = (2 \times 16.0) = 32.0$.
    - Pressure = 200 kPa = 200000 Pa.
    - Moles O_2 = Mass ÷ M_r = 7 ÷ 32.0 = 0.219 mol (3 significant figures).
    - Rearranging the Ideal Gas Law, we get Volume = Moles × Temperature × R ÷ Pressure.
    - Volume O_2 = 0.219 × 350 × 8.314 ÷ 200000 = 0.00318 m³ = 3.18 dm³ (3 significant figures).
13. Solution:
    - $M_r\ O_2 = (2 \times 16.0) = 32.0$.
    - Temperature = -40°C = (-40 + 273.15) = 233.15K.
    - Pressure = 100 kPa = 100000 Pa.
    - Moles O_2 = Mass ÷ M_r = 7 ÷ 32.0 = 0.219 mol (3 significant figures).
    - Rearranging the Ideal Gas Law, we get Volume = Moles × Temperature × R ÷ Pressure.
    - Volume O_2 = 0.219 × 233.15 × 8.314 ÷ 100000 = 0.00424 m³ = 4.24 dm³ (3 significant figures).
14. Solution:
    - Pressure = 100 kPa = 100000 Pa.
    - Rearranging the Ideal Gas Law, we get Volume = Moles × Temperature × R ÷ Pressure.
    - Volume HI = 0.5 × 300 × 8.314 ÷ 100000 = 0.0125 m³ = 12.5 dm³ (3 significant figures).
15. Solution:
    - $M_r\ N_2 = (2 \times 14.0) = 28.0$.
    - Pressure = 100 kPa = 100000 Pa.

- Volume = 32 dm³ = 0.032 m³.
- Moles N_2 = Mass ÷ M_r = 20 ÷ 28.0 = 0.714 mol (3 significant figures).
- Rearranging the Ideal Gas Law, we get Temperature = (Pressure × Volume) ÷ (Moles × R).
- Temperature = (100000 × 0.032) ÷ (0.714 × 8.314) = 539K = 266°C (3 significant figures).

16. Solution:
    - M_r Ca(NO₃)₂ = (1 × 40.1) + (2 × 14.0) + (6 × 16.0) =164.1.
    - Moles Ca(NO₃)₂ = Mass ÷ M_r = 10 ÷ 164.1 = 0.0609 mol (3 significant figures).
    - The ratio of Ca(NO₃)₂ : Gaseous product molecules is 2 : (4 + 1) = 2 : 5, so Moles of Gaseous Products = 0.0609 × (5 ÷ 2) = 0.152 mol (3 significant figures).
    - Volume of Gaseous Products = Moles × 24 = 0.152 × 24 = 3.66 dm³ (3 significant figures).

17. Solution:
    - Volume HCl = 1300 cm³ = 1.300 dm³ = 0.001300 m³.
    - Pressure = 100 kPa = 100000 Pa.
    - Using the Ideal Gas Law, Moles HCl = (Pressure × Volume) ÷ (R × Temperature) = (100000 × 0.001300) ÷ (8.314 × 320) = 0.0489 mol (3 significant figures).
    - Volume H_2O = 5000 cm³ = 5 dm³.
    - Concentration HCl (mol dm⁻³) = Moles ÷ Volume = 0.0489 ÷ 5 = 0.00977 mol dm⁻³ (3 significant figures).

18. Solution:
    - Volume Unknown Gas = 15 liter = 15 dm³.
    - Moles Unknown Gas = Volume ÷ 24 = 15 ÷ 24 = 0.625 mol.
    - M_r Unknown Gas = Mass ÷ Moles = 27.5 ÷ 0.625 = 44.
    - M_r H_2 = (2 × 1.0) = 2.0.
    - M_r He = (1 × 4.0) = 4.0.
    - M_r N_2 = (2 × 14.0) = 28.0.
    - M_r O_2 = (2 × 16.0) = 32.0.
    - M_r CO_2 = (1 × 12.0) + (2 × 16.0) = 44.0.
    - The only candidate gas that has the same M_r as that of the actual gas is carbon dioxide (CO_2), so therefore the gas is carbon dioxide.

19. Solution:
    - The maximum volume would be when all the aluminium sulfide reacts (aluminium sulfide is the limiting reactant).
    - M_r Al_2S_3 = (2 × 27.0) + (3 × 32.1) = 150.3.
    - Moles Al_2S_3 = Mass ÷ M_r = 10 ÷ 150.3 = 0.0665 mol (3 significant figures).
    - The ratio of Al_2S_3 : H_2S is 1 : 3, therefore moles of H_2S = 0.0665 × 3 = 0.200 mol (3 significant figures).
    - Volume of H_2S = Moles × 24 = 0.200 × 24 = 4.79 dm³ (3 significant figures).

20. Solution:
    - Temperature = 450°C = 723.15K.
    - Pressure = 21 MPa = 21000 kPa = 21000000 Pa.

- Using the Ideal Gas Law, Moles Gas = (Pressure × Volume) ÷ (R × Temperature) = (21000000 × 800) ÷ (8.314 × 723.15) = 2,790,000 mol (3 significant figures).

Chapter 5: Atom Economy

Atom economy (also known as **atom efficiency**) is measure of how efficient a chemical process is in converting the reactants into the desired products. It might be considered a measure of how green the chemical process is – as a high atom economy indicates that the process produces little or no waste products.
- A chemical reaction (or a chemical process) that produces only the desired product (or desired products) has an atom economy of 100%.
- A chemical reaction (or a chemical process) that produces only unwanted products has an atom economy of 0%.
- A chemical reaction (or a chemical process) that produces both desired and unwanted products would have an atom economy somewhere between 0% and 100%.

Calculating the Atom Economy of a Reaction

To calculate the atom of economy of any chemical reaction, simply add up the relative masses of the desired products (taking into account the amount of each product present), add up the relative masses of all the products (again, taking into account the amount of each product present), divide the former by the latter and multiply by 100. Note: because of the law of conservation of mass, if you wish, you can use the total relative mass of all reactants instead of the total relative mass of all the products.

Example: What is the atom economy for the production of ethanol by the fermentation of glucose: $C_6H_{12}O_6 \rightarrow 2CH_3CH_2OH + 2CO_2$ (assume carbon dioxide is a waste product)?

Solution:
- M_r of CH_3CH_2OH = (2 × 12.0) + (6 × 1.0) + (1 × 16.0) = 46.0.
- M_r of CO_2 = (1 × 12.0) + (2 × 16.0) = 44.0.
- Relative mass of desired products = (2 × 46.0) = 92.0.
- Relative mass of all products = (2 × 46.0) + (2 × 44.0) = 180.0.
- Atom economy = (Desired Products ÷ All Products) × 100 = (92.0 ÷ 180.0) × 100 = 51.1% (3 significant figures).

Alternate solution:
- M_r of CH_3CH_2OH = (2 × 12.0) + (6 × 1.0) + (1 × 16.0) = 46.0.
- M_r of $C_6H_{12}O_6$ (6 × 12.0) + (12 × 1.0) + (6 × 16.0) = 180.0.
- Relative mass of desired products = (2 × 46.0) = 92.0.
- Relative mass of all reactants = (1 × 180.0) = 180.0.
- Atom economy = (Desired Products ÷ All Reactants) × 100 = (92.0 ÷ 180.0) × 100 = 51.1% (3 significant figures).

Calculating the Atom Economy of a Chemical Process Involving Multiple Steps

Many chemical processes involve multiple steps, with the only some of products produced in the earlier steps being used in later steps. For example, consider the following series of reactions:
- Step 1: A + B → C + D
- Step 2: C + E → F + G
- Step 3: F + H → J + K
- Step 4: J + L → M + N

If you look carefully, you will see that although C is produced in step 1, it is then consumed in step 2. Likewise, F is produced in step 2, but then is consumed in step 3. And J is produced in step 3, but then is consumed in step 4.

We can divide the chemicals in the reaction into three categories:
- Reactants that we must supply: A, B, E, H and L.
- Products that are produced in one step but consumed in a later step: C, F and J.
- Products left at the end of the series of the reactions: D, G, K, M and N.

When calculating atom economy, we can simply disregard the products that are later consumed (C, F and J). The atom economy can be calculated by dividing the relative mass of the desired product(s) by the relative mass of the reactants supplied (A, B, E, H and L) and multiplying by 100, or by dividing the relative mass of the desired product(s) by the relative mass of all products left at the end of the series (D, G, K, M and N) and multiplying by 100.

So, if M was the only desired product:
- Atom economy = (M ÷ (A + B + E + H + L)) × 100.
- Atom economy = (M ÷ (D + G + K + M + N)) × 100.

Lastly, one important consideration when writing a multistep process and calculating its atom economy, is that the quantities of products created in one stage and subsequently consumed must match. For example, consider this two-step process:
- Step 1: P + Q → R + S
- Step 2: 2R + T → U + V

Notice how R is produced in step 1, but 2R is consumed in step 2 – they do quantities produced and consumed do **not** match! So, we should write the chemical equations like this:
- Step 1: 2P + 2Q → 2R + 2S
- Step 2: 2R + T → U + V

Then, as before, we can divide the chemicals in the reaction into three categories:
- Reactants that we must supply: 2P, 2Q and T.
- Products that are produced in one step but consumed in a later step: 2R.

- Products left at the end of the series of the reactions: 2S, U and V.

So, if U was the only desired product:
- Atom economy = (U ÷ (2P + 2Q + T)) × 100.
- Atom economy = (U ÷ (2S + U + V)) × 100.

Example: Calculate the atom economy for the production of phenol (C_6H_5OH) from benzene (C_6H_6) using a two-step synthesis, step 1 being the chlorination of benzene in the presence of anhydrous aluminium chloride, and step 2 being the reacting the chlorobenzene with steam (H_2O) in the presence of calcium phosphate. The chemical reactions are:
- Step 1: $C_6H_6 + Cl_2 \rightarrow C_6H_5Cl + HCl$
- Step 2: $C_6H_5Cl + H_2O \rightarrow C_6H_5OH + HCl$

Solution:
- All products left at the end = 2HCl and C_6H_5OH
- Desired product = C_6H_5OH.
- M_r HCl = (1 × 1.0) + (1 × 35.5) = 36.5.
- M_r C_6H_5OH = (6 × 12.0) + (6 × 1.0) + (1 × 16.0) = 94.0.
- Relative mass of all products = (2 × 36.5) + 94.0 = 167.0.
- Relative mass of desired product = 94.0.
- Atom economy = (Desired Products ÷ All Products) × 100 = (94.0 ÷ 167.0) × 100 = 56.3% (3 significant figures).

Alternative solution:
- All reactants to be supplied = C_6H_6, Cl_2 and H_2O.
- Desired product = C_6H_5OH
- M_r C_6H_6 = (6 × 12.0) + (6 × 1.0) = 78.0
- M_r Cl_2 = (2 × 35.5) = 71.0
- M_r H_2O = (2 × 1.0) + (1 × 16.0) = 18.0.
- M_r C_6H_5OH = (6 × 12.0) + (6 × 1.0) + (1 × 16.0) = 94.0.
- Relative mass of desired product = 94.0.
- Relative mass of reactants to be supplied = 78.0 + 71.0 + 18.0 = 167.0.
- Atom economy = (Desired Products ÷ Reactants To Be Supplied) × 100 = (94.0 ÷ 167.0) × 100 = 56.3% (3 significant figures).

Improving Atom Economy

As atom economy is one metric for how green a chemical reaction or process is, chemists will often design chemical processes so as to maximize atom economy, especially for chemical processes performed on a large scale.

You Can Do Chemistry: Moles & Stoichiometry

There are a number of ways that this can be done:
- Choosing chemical processes that produce few or no unwanted products (known as **by-products** or **secondary products**). For example, addition reactions (where two chemicals combine to produce a single chemical with no other product) would have an atom economy of 100%.
- If a chemical process produces two or more products, finding a use for all (or as many as possible) products.

Of course, atom economy is **not** the only factor that chemists need to consider when designing chemical processes. Although a particular reaction may have high atom economy, it might produce a poor yield (see Chapter 6), require expensive or hard-to-obtain reactants or catalysts, require large amounts of energy, require high pressures (generating high pressures is expensive), etc.

Questions

1. What is the atom economy for the production of iron (Fe) by reducing iron (III) oxide using carbon (assume all products except iron are unwanted byproducts)? The chemical equation for this reaction is: $Fe_2O_3(s) + 3C(s) \rightarrow 2Fe(l) + 3CO(g)$.

2. What is the atom economy for the production of copper (Cu) by reducing copper (II) oxide using carbon (assume all products except copper are unwanted byproducts)? The chemical equation for this reaction is $2CuO(s) + C(s) \rightarrow 2Cu(l) + CO_2(g)$

3. Nitric acid (HNO_3) is produced in industry by the Ostwald process which involves the oxidation of ammonia over a platinum catalyst at 700-850°C. What is the atom economy for this process (assume all products except nitric acid are unwanted byproducts)? The overall chemical equation for the process is $NH_3 + 2O_2 \rightarrow HNO_3 + H_2O$.

4. Hydrazine (H_2N-NH_2) can be synthesized by oxidizing urea (($H_2N)_2C=O$). What is the atom economy for this process (assume all products except hydrazine are unwanted byproducts)? The overall chemical equation for this process is $(H_2N)_2C=O + NaOCl + 2NaOH \rightarrow H_2N-NH_2 + H_2O + NaCl + Na_2CO_3$.

5. Hydrazine (H_2N-NH_2) can also be synthesized using the Olin Raschig process. What is the atom economy for this process (assume all products except hydrazine are unwanted byproducts)? The chemical equations for the process are:
- Step 1: $NaOCl + NH_3 \rightarrow NH_2Cl + NaOH$.
- Step 2: $NH_2Cl + NH_3 + NaOH \rightarrow H_2N-NH_2 + NaCl + H_2O$

6. Hydrazine (H_2N-NH_2) can also be synthesized using the peroxide process (also known as the Pechiney-Ugine-Kuhlmann process and several other names). What is the atom economy for this process (assume all products except hydrazine are unwanted byproducts)? The overall chemical equation for this process is $2NH_3 + H_2O_2 \rightarrow H_2N-NH_2 + 2H_2O$.

7. The Hunter process was earliest commercial process used to prepare titanium (Ti) from titanium (IV) chloride ($TiCl_4$). What is the atom economy for this process (assume all products except titanium

are unwanted byproducts)? The chemical equation for the Hunter process is 4Na + TiCl$_4$ → 4NaCl + Ti.

8. In industry today, the Kroll process is used to prepare titanium (Ti) from titanium (IV) chloride (TiCl$_4$). What is the atom economy for this process (assume all products except titanium are unwanted byproducts)? The chemical equation for the Kroll process is 2Mg + TiCl$_4$ → 2MgCl$_2$ + Ti.

9. One of the most common ores of titanium is rutile which primarily consists of titanium (IV) oxide. Titanium metal can be obtained by a two-step process, which involves the production of titanium (IV) chloride followed by the Kroll process. What is the atom economy for this process (assume all products except titanium are unwanted byproducts)? The chemical equations for the overall process are:

- Step 1: TiO$_2$ + C + 2Cl$_2$ → TiCl$_4$ + 2CO
- Step 2 (Kroll process): 2Mg + TiCl$_4$ → 2MgCl$_2$ + Ti

10. Another common ore of titanium is ilmenite (also known as menaccanite) which consists of iron titanium oxide (FeTiO$_3$). Titanium metal can be obtained by a three-step process, which involves the reduction of the iron, followed by the production of titanium (IV) chloride, finally followed by the Kroll process. What is the atom economy for this process (assume all products except titanium are unwanted byproducts)? The chemical equations for the overall process are:

- Step 1: FeTiO$_3$ + C → Fe + TiO$_2$ + CO
- Step 2: TiO$_2$ + C + 2Cl$_2$ → TiCl$_4$ + 2CO
- Step 3 (Kroll process): 2Mg + TiCl$_4$ → 2MgCl$_2$ + Ti

Ilemnite ore:

Answers to Chapter 5 Questions

1. Solution:
    - A_r of Fe = 55.8.
    - M_r of CO = (1 × 12.0) + (1 × 16.0) = 28.0.
    - Relative mass of desired products = (2 × 55.8) = 111.6.
    - Relative mass of all products = (2 × 55.8) + (3 × 28.0) = 195.6.
    - Atom economy = (Desired Products ÷ All Products) × 100 = (111.6 ÷ 195.6) × 100 = 57.1% (3 significant figures).

2. Solution:
    - A_r of Cu = 63.5.
    - M_r of CO_2 = (1 × 12.0) + (2 × 16.0) = 44.0.
    - Relative mass of desired products = (2 × 63.5) = 127.0.
    - Relative mass of all products = (2 × 63.5) + (1 × 44.0) = 171.0.
    - Atom economy = (Desired Products ÷ All Products) × 100 = (127.0 ÷ 171.0) × 100 = 74.3% (3 significant figures).

3. Solution:
    - M_r of HNO_3 = (1 × 1.0) + (1 × 14.0) + (3 × 16.0) = 63.0.
    - M_r of H_2O = (1 × 1.0) + (1 × 16.0) = 18.0.
    - Relative mass of desired products = (1 × 63.0) = 63.0.
    - Relative mass of all products = (1 × 63.0) + (1 × 18.0) = 81.0.
    - Atom economy = (Desired Products ÷ All Products) × 100 = (63.0 ÷ 81.0) × 100 = 77.8% (3 significant figures).

4. Solution:
    - M_r of $H_2N\text{-}NH_2$ = (4 × 1.0) + (2 × 14.0) = 32.0.
    - M_r of H_2O = (1 × 1.0) + (1 × 16.0) = 18.0.
    - M_r of NaCl = (1 × 23.0) + (1 × 35.5) = 58.5.
    - M_r of Na_2CO_3 = (2 × 23.0) + (1 × 12.0) + (3 × 16.0) = 106.0.
    - Relative mass of desired products = (1 × 32.0) = 32.0.
    - Relative mass of all products = (1 × 32.0) + (1 × 18.0) + (1 × 58.5) + (1 × 106.0) = 214.5
    - Atom economy = (Desired Products ÷ All Products) × 100 = (32.0 ÷ 214.5) × 100 = 14.9% (3 significant figures).

5. Solution:
    - All products left at the end = $H_2N\text{-}NH_2$, NaCl and H_2O
    - Desired product = $H_2N\text{-}NH_2$
    - M_r of $H_2N\text{-}NH_2$ = (4 × 1.0) + (2 × 14.0) = 32.0.
    - M_r of NaCl = (1 × 23.0) + (1 × 35.5) = 58.5.
    - M_r of H_2O = (1 × 1.0) + (1 × 16.0) = 18.0.
    - Relative mass of all products = 32.0 + 58.5 + 18.0 = 108.5.
    - Relative mass of desired product = 32.0.

- Atom economy = (Desired Products ÷ All Products) × 100 = (32.0 ÷ 108.5) × 100 = 29.5% (3 significant figures).

6. Solution:
   - M_r of H_2N-NH_2 = (4 × 1.0) + (2 × 14.0) = 32.0.
   - M_r of H_2O = (1 × 1.0) + (1 × 16.0) = 18.0.
   - Relative mass of desired products = (1 × 32.0) = 32.0.
   - Relative mass of all products = (1 × 32.0) + (2 × 18.0) = 68.0.
   - Atom economy = (Desired Products ÷ All Products) × 100 = (32.0 ÷ 68.0) × 100 = 47.1% (3 significant figures).

7. Solution:
   - M_r of NaCl = (1 × 23.0) + (1 × 35.5) = 58.5.
   - A_r of Ti = 47.9.
   - Relative mass of desired products = (1 × 47.9) = 47.9.
   - Relative mass of all products = (4 × 58.5) + (1 × 47.9) = 281.9.
   - Atom economy = (Desired Products ÷ All Products) × 100 = (47.9 ÷ 281.9) × 100 = 17.0% (3 significant figures).

8. Solution:
   - M_r of $MgCl_2$ = (1 × 24.3) + (2 × 35.5) = 95.3.
   - A_r of Ti = 47.9.
   - Relative mass of desired products = (1 × 47.9) = 47.9.
   - Relative mass of all products = (2 × 95.3) + (1 × 47.9) = 238.5.
   - Atom economy = (Desired Products ÷ All Products) × 100 = (47.9 ÷ 238.5) × 100 = 20.1% (3 significant figures).

9. Solution:
   - All products left at the end = 2CO, $2MgCl_2$ and Ti.
   - Desired product = Ti
   - M_r of CO = (1 × 12.0) + (1 × 16.0) = 28.0.
   - M_r of $MgCl_2$ = (1 × 24.3) + (2 × 35.5) = 95.3.
   - A_r of Ti = 47.9.
   - Relative mass of all products = (2 × 28.0) + (2 × 95.3) + (1 × 47.9) = 294.5.
   - Relative mass of desired product = 47.9.
   - Atom economy = (Desired Products ÷ All Products) × 100 = (47.9 ÷ 294.5) × 100 = 16.3% (3 significant figures).

10. Solution:
    - All products left at the end = Fe, CO, 2CO, $2MgCl_2$ and Ti.
    - Desired product = Ti
    - A_r of Fe = 55.8.
    - M_r of CO = (1 × 12.0) + (1 × 16.0) = 28.0.
    - M_r of $MgCl_2$ = (1 × 24.3) + (2 × 35.5) = 95.3.
    - A_r of Ti = 47.9.

- Relative mass of all products = (1 × 55.8) + (1 × 28.0) + (2 × 28.0) + (2 × 95.3) + (1 × 47.9) = 378.3.
- Relative mass of desired product = 47.9.
- Atom economy = (Desired Products ÷ All Products) × 100 = (47.9 ÷ 378.3) × 100 = 12.7% (3 significant figures).

Chapter 6: Chemical Yield

Chemical yield (also known as **reaction yield**) is the amount of product produced in a chemical reaction or process. There are several ways to express yield.

Absolute Yield and Molar Yield

Absolute yield is simply the amount of product produced measured by mass or in moles. If expressed in moles, it is also known as **molar yield**. When contrasting absolute or molar yield to what we might have theoretically expected, it is known as **actual yield**.

Example: The chemical reaction of ethene ($CH_2=CH_2$) with steam (H_2O) is used to produce ethanol (CH_3CH_2OH), using the following chemical reaction: $CH_2=CH_2 + H_2O \rightarrow CH_3CH_2OH$.

A total of 80 g of ethanol are produced and collected.
- Therefore, the absolute yield of ethanol is 80 g.

We can calculate the molar yield:
- M_r of ethanol = (2 × 12.0) + (6 × 1.0) + (1 × 16.0) = 46.0.
- Molar yield = 80 ÷ 46.0 = 1.74 moles (3 significant figures).

Theoretical Yield

Theoretical yield is the amount of product predicted by a stoichiometric calculation based on the amount of each the reactants present. For the purposes of this calculation, it is assumed that only one reaction occurs and that the limiting reactant reacts completely.

Example: 25 g of calcium carbonate (CaCO3) chips are placed in a large conical flask. 120 cm^3 of 1.2 mol dm^{-3} hydrochloric acid (HCl) is added to the flask, and any gases emitted is collected using a gas syringe. The following reaction takes place: $CaCO_3(s) + 2HCl(aq) \rightarrow CaCl_2(aq) + H_2O(l) + CO_2(g)$. What is the theoretical yield of carbon dioxide?

Solution:
- M_r of $CaCO_3$ = (1 × 40.1) + (1 × 12.0) + (3 × 16.0) = 100.1.
- Moles $CaCO_3$ = 25 ÷ 100.1 = 0.250 mol (3 significant figures).
- Moles HCl = 1.2 × 0.120 = 0.144 mol (3 significant figures).
- Since the ratio of $CaCO_3$: HCl in the chemical equation is 1 : 2, 0.250 moles of $CaCO_3$ could react with up to 2 × 0.250 = 0.500 moles of HCl (there is **not** enough HCl), whereas 0.144 moles of HCl can react with 0.144 ÷ 2 = 0.072 moles of $CaCO_3$ (there is more than enough $CaCO_3$). So HCl is the limiting reactant, and $CaCO_3$ is the excess reactant.
- Since the ratio of HCl : CO_2 in the chemical equation is 2 : 1, the theoretical yield of CO_2 is 0.144 ÷ 2 = 0.072 moles.

- M_r of CO_2 = (1 × 12.0) + (2 × 16.0) = 44.0.
- Theoretical yield of CO_2 in grams = Moles × M_r = 0.072 × 44.0 = 3.17 g (3 significant figures).

Actual Yield is Always Less Than Theoretical Yield

The theoretical yield expresses the maximum possible yield for a chemical process, but the actual yield achieved by performing the process will always be less.

There are several reasons for this:
- The reaction may **not** be complete – **not** all reactants may be converted to products.
- The reaction may be reversible (both the forward and a reverse reaction occur), and a dynamic equilibrium might be established between these forward and backward reactions, meaning that a mix of both products and reactants will be present.
- Additional reactions might occur in addition to the reaction(s) that we aimed for, meaning that some of reactants might get converted into undesired side products rather than the desired product(s).
- Losses of material may occur during the separation and purification of the desired product from the reaction mixture. For example, in the laboratory, transferring between beakers might lose some product, some product might be accidentally washed way during filtration, or gases may escape from laboratory equipment which is **not** perfectly airtight.
- The starting reactants may contain impurities which do **not** react to produce the desired product(s).

Fractional Yield and Percentage Yield

Fractional yield and percentage yield (also known as relative yield) measure how successful a chemical process was at actually producing the products expected by the theoretical reaction(s).
- Fractional yield is calculated by dividing the actual yield achieved by the theoretical yield. Any units can be used for the actual and theoretical yields (grams, moles, gas volume at a specified temperature and pressure) provided that the same units are used for both.
- Percentage yield is calculated by multiplying the fractional yield by 100.

Example: 25 cm³ of 0.8 mol dm⁻³ silver nitrate ($AgNO_3$) is mixed with excess hydrochloric acid solution (HCl). The following chemical reaction takes place generating a precipitate of silver chloride (AgCl): $AgNO_3(aq)$ + HCl(aq) → AgCl(s) + $HNO_3(aq)$. What is the theoretical yield of silver nitrate? If a student achieves an actual yield of 2.31 g of silver nitrate, what was her percentage yield?
- Moles $AgNO_3$ = 0.8 × 0.025 = 0.02 mol.
- Ratio in chemical equation of $AgNO_3$: AgCl is 1 : 1, therefore theoretical molar yield of AgCl = 0.02 mol.
- M_r of AgCl = (1 × 107.8) + (1 × 35.5) = 143.3.
- Theoretical yield of AgCl = Moles × M_r = 0.02 × 143.3 = 2.866 g.

- Actual yield = 2.31 g.
- Fractional yield = (Actual Yield ÷ Theoretical Yield) = 0.806 (3 significant figures).
- Percentage yield = 0.806 × 100 = 80.6% (3 significant figures).

Example: An excess of calcium carbonate ($CaCO_3$) is reacted with 10 cm^3 of 0.6 mol dm^{-3} hydrochloric acid (HCl) under r.t.p conditions, the following reaction takes place: $CaCO_3(s) + 2HCl(aq)$ → $CaCl_2(aq) + H_2O(l) + CO_2(g)$. What is the theoretical yield of carbon dioxide (as a volume of gas)? If a student achieves an actual yield of 37.2 cm^3 of carbon dioxide, what was the percentage yield?
- Moles HCl = 0.6 × 0.010 = 0.006 mol.
- Ratio in chemical equation of HCl : CO_2 is 2 : 1, therefore theoretical molar yield of CO_2 = 0.006 ÷ 2 = 0.003 mol.
- Theoretical yield of CO_2 (volume at r.tp) = 0.003 × 24 = 0.072 dm^3 = 72 cm^3.
- Actual yield = 37.2 cm^3.
- Fractional yield = (Actual Yield ÷ Theoretical Yield) = 37.2 ÷ 72 = 0.517 (3 significant figures).
- Percentage Yield = 0.517 × 100 = 51.7% (3 significant figures).

The Difference Between Atom Economy and Percentage Yield

Percentage yield is very different issue from atom economy (see Chapter 5):
- Percentage yield measures how much of the products were actually obtained compared to what was theoretically possible.
- Atom economy measures what proportion of all the products of the principle reaction process (remember in real processes, unwanted reactions may take place as well) would be desirable products.

Example: Ammonia (NH_3) is manufactured by reacting nitrogen (N_2) and hydrogen (H_2) gases using the Haber process, which utilizes the following chemical reaction: $N_2(g) + 3H_2(g) \rightleftharpoons 2NH_3(g)$.
- The atom economy of this reaction is 100%, since only the desired product is produced.
- The percentage yield achieved in industry is typically between 10% and 20% because under the conditions used in industry, the equilibrium of the reversible reaction lies to the left and only a small proportion of the reactants are converted into ammonia.

Example: Consider the reaction of propene ($CH_3CH=CH_2$) with hydrobromic acid (HBr) to produce 2-bromopropane ($CH_3CHBrCH_3$): $CH_3CH=CH_2 + HBr$ → $CH_3CHBrCH_3$.
- The atom economy of this reaction is 100%, since according to the chemical equation, only the desired product is produced.
- The percentage yield when attempting this reaction will always be well under 100%. This is because, although some of the reactants will undergo the principle reaction described above, a proportion of the reactants always undergo an alternative reaction which produces 1-bromopropane ($CH_3CH_2CH_2Br$) instead: $CH_3CH=CH_2 + HBr$ → $CH_3CH_2CH_2Br$.

How to Maximize Percentage Yield

When we review the reasons why actual yield is always less than theoretical yield, it turns out that there may be things that we can do maximize actual yield, and thus increase the percentage yield of any chemical process:

- One reason for a reduction in yield is that the reaction may **not** be complete, and some of the reactants may **not** have been converted into products. Allowing more time for the reaction to complete may help with this. Additionally, performing the reaction at a higher temperature, or a higher pressure or concentration, or with a catalyst, may speed up the reaction, allowing a higher yield in less time.
- In the case of reversible reactions (like the Haber process), a dynamic equilibrium might be established between the forward and backward reactions, meaning that a mix of both products and reactants are present. Le Chatelier's principle teaches us that if we change the conditions of a reversible reaction, the position of equilibrium moves to oppose the change. A reversible reaction that in the forward direction reduces the amount of gas particles present, produces the highest yields at high pressure (and a reversible reaction that increases the number gas particles will produce the highest yields at low pressure). A reversible reaction that is exothermic in the forward direction will produce the highest yields at low temperature (and one that is endothermic in the forward direction will produce the highest yields at high temperature).
- If additional reactions occur as well as the one that want, and these additional reactions produce unwanted side products that reduce the yield of the desired product, it may be possible to find catalyst specific to the preferred reaction, or otherwise change the conditions to favor the preferred reaction.
- Care should be taken during the any separation and purification stages, so as to lose as little as possible of the desired product(s). For example, repeatedly transferring chemicals between containers, excessive washing of a solid, or the equipment **not** being completely airtight in the case of a gaseous product, are all common reasons for reductions in yield in the laboratory.
- If the starting reactants contain impurities, these also will reduce the yield, using purer starting materials will therefore result in higher yields.

Over and above this, there may be more than one method to manufacture or synthesize a desired chemical, and therefore there may be a choice of several different chemical processes and routes. Each different route will have its own advantages and disadvantages including the ease of performing the steps, the availability of starting materials and catalysts, the number of steps necessary, etc., and of course the final yield of the desired product.

Of course, maximizing yield is **not** the only consideration when performing chemical process. For an example, a process which requires exotic catalysts, or very high temperatures or pressures, might **not** be achievable (or might **not** be economic in industry). Similarly, even if a low temperature maximizes the yield of a reversible reaction, it might be more practical to use a hotter compromise temperature which produces a lower yield, but in less time (as rates of reaction speed up as temperature increases).

Questions

1. John heats up 7 g of calcium carbonate ($CaCO_3$) expecting it to undergo thermal decomposition according to the following chemical equation: $CaCO_3(s) \rightarrow CaO(s) + CO_2(g)$. What is his theoretical yield of calcium oxide (CaO)?

2. Sally also heats up some of calcium carbonate ($CaCO_3$) expecting it to undergo thermal decomposition according to the following chemical equation: $CaCO_3(s) \rightarrow CaO(s) + CO_2(g)$. Sally predicts that she will produce 6.5 g of calcium oxide (CaO), but only produces 5.9 g. What is her percentage yield?

3. 17 cm³ of 0.8 mol dm⁻³ sodium sulfide (Na_2S) solution are mixed with 25 cm³ of 1.2 mol dm⁻³ silver nitrate ($AgNO_3$) solution. A precipitate of silver sulfide (Ag_2S) is formed by the following chemical reaction: $Na_2S(aq) + 2AgNO_3(aq) \rightarrow NaNO_3(aq) + Ag_2S(s)$. What is the theoretical yield of silver sulfide? If the actual yield of silver sulfide is 3.11 g, what is the percentage yield?

4. In a chemical plant, 2300 tons of hydrogen (H_2) are reacted with excess nitrogen (N_2) to produce 2000 tons of ammonia (NH_3), according to the following chemical equation: $N_2 + 3H_2 \rightleftharpoons 2NH_3(g)$. What is the percentage yield?

5. A rival chemical plant also produces ammonia (NH_3) from nitrogen (N_2) and hydrogen (H_2) using the same chemical reaction: $N_2 + 3H_2 \rightleftharpoons 2NH_3(g)$. This plant says it can produce 2000 tons of ammonia using a chemical process with a percentage yield of 20%, how many tons of nitrogen and hydrogen would be needed if this claim were true?

6. 0.2 g of calcium nitrate ($Ca(NO_3)_2$) is heated and undergoes thermal decomposition according to the following chemical equation: $2Ca(NO_3)_2(s) \rightarrow 2CaO(s) + 4NO_2(g) + O_2(g)$. The gas is collected in a gas syringe and has a volume of 48 cm³ at r.t.p. What is the percentage yield of the gas?

7. In the presence of a manganese (IV) oxide catalyst, potassium chlorate ($KClO_3$) can be heated in a test tube to produce oxygen (O_2) gas according to the following chemical equation: $2KClO_3(s) \rightarrow 2KCl(s) + 3O_2(g)$. If 1 g of potassium chlorate is used what is the theoretical yield of oxygen (as a volume) at r.t.p? If the actual volume of oxygen collected is 221 cm³ what is the actual yield?

8. 25 cm³ of 1.1 mol dm⁻³ hydrochloric acid (HCl) is mixed with 2 g of manganese (IV) oxide (MnO_2). The following chemical reaction occurs: $MnO_2(s) + 4HCl(aq) \rightarrow MnCl_2(aq) + 2H_2O(l) + Cl_2(g)$. If the percentage yield is 38%, what volume of chlorine gas (Cl_2) is produced at r.t.p?

9. Magnesium nitrate ($Mg(NO_3)_2$) can be prepared by reacting nitric acid (HNO_3) with magnesium oxide (MgO) according to the following chemical reaction: $2HNO_3(aq) + MgO(s) \rightarrow Mg(NO_3)_2(aq) + H_2O(l)$. John performed this reaction in the laboratory using 200 cm³ of nitric acid and excess magnesium oxide. If he produced 3.9 g of magnesium nitrate, and his percentage yield was 79%, what was the concentration of the nitric acid?

10. Sally also prepared magnesium nitrate ($Mg(NO_3)_2$) by reacting nitric acid (HNO_3) with magnesium oxide (MgO) according to the following chemical reaction: $2HNO_3(aq) + MgO(s) \rightarrow Mg(NO_3)_2(aq) +$

H₂O(l). Sally used 1.1 mol dm⁻³ of nitric acid and excess magnesium oxide. If she produced 4.3 g of magnesium nitrate, and her percentage yield was 81%, what volume of nitric acid did she use?

Answers to Chapter 6 Questions

1. Solution:
    - M_r CaCO$_3$ = (1 × 40.1) + (1 × 12.0) + (3 × 16.0) = 100.1.
    - M_r CaO = (1 × 40.1) + (1 × 16.0) = 56.1.
    - Moles CaCO$_3$ = 7 ÷ 100.1 = 0.0699 mol (3 significant figures).
    - Ratio CaCO$_3$: CaO is 1:1, therefore theoretical moles CaO = 0.0699.
    - Theoretical yield CaO = 0.0699 × 56.1 = 3.92 g (3 significant figures).

2. Solution:
    - Percentage Yield = (Actual Yield ÷ Theoretical Yield) × 100 = (5.9 ÷ 6.5) = 90.8% (3 significant figures).

3. Solution:
    - Moles Na$_2$S = 0.017 × 0.8 = 0.0136 mol.
    - Moles AgNO$_3$ = 0.025 × 1.2 = 0.03 mol.
    - M_r Ag$_2$S = (2 × 107.9) + (1 × 32.1) = 247.9.
    - The chemical equation indicates a ratio of Na$_2$S : AgNO$_3$ of 1 : 2, but as there is more than twice as much AgNO$_3$, Na$_2$S is the limiting reactant, and AgNO$_3$ is in excess.
    - The chemical equation indicates a ratio of Na$_2$S : Ag$_2$S of 1 : 1, so the theoretical molar yield of Ag$_2$S is 0.0136 mol.
    - Theoretical absolute yield of Ag$_2$S = 0.0136 × 247.9 = 3.37 g (3 significant figures).
    - Percentage Yield = (Actual Yield ÷ Theoretical Yield) × 100 = (3.11 ÷ 3.37) = 92.2% (3 significant figures).

4. Solution:
    - 2300 tons = 2300000000 g.
    - 2000 tons = 2000000000 g.
    - M_r H$_2$ = (2 × 1.0) = 2.0.
    - M_r NH$_3$ = (1 × 14.0) + (3 × 1.0) = 17.0.
    - Moles H$_2$ = 2300000000 ÷ 2.0 = 1150000000 mol.
    - The chemical equation indicates a ratio of H$_2$: NH$_3$ of 3 : 2, therefore the theoretical molar yield of NH$_3$ = 1150000000 × (2 ÷ 3) = 766666667 mol.
    - Actual yield of NH$_3$ = 2000000000 ÷ 17 = 117647059 mol.
    - Percentage Yield = (Actual Yield ÷ Theoretical Yield) × 100 = (117647059 ÷ 766666667) × 100 = 15.3% (3 significant figures).

5. Solution:
    - Rearranging Percentage Yield = (Actual Yield ÷ Theoretical Yield) × 100, we get Theoretical Yield = (Actual Yield ÷ Percentage Yield) × 100 = (2000 ÷ 20) × 100 = 10000 tons.
    - 10000 tons = 10000000000 g.
    - M_r NH$_3$ = (1 × 14.0) + (3 × 1.0) = 17.0.
    - M_r N$_2$ = (2 × 14.0) = 28.0.
    - M_r H$_2$ = (2 × 1.0) = 2.0.
    - Theoretical molar yield of NH$_3$ = 10000000000 ÷ 17.0 = 588235294 mol

- The chemical equation indicates a ratio of $NH_3 : H_2$: of 2 : 3, therefore the theoretical molar amount of H_2 = 588235294 × (3 ÷ 2) = 882352941 mol (if H_2 is the limiting reactant).
- Therefore, if H_2 is the limiting reactant, mass H_2 = 882352941 × 2.0 = 1764705882 g = 1760 tons (3 significant figures), but more H_2 would be needed if it is in excess.
- The chemical equation indicates a ratio of $NH_3 : N_2$: of 2 : 1, therefore the theoretical molar amount of N_2 = 588235294 ÷ 2 = 294117647 mol (if N_2 is the limiting reactant).
- Therefore, if N_2 is the limiting reactant, mass N_2 = 294117647 × 28.0 = 8235294118 g = 8240 tons (3 significant figures), but more N_2 would be needed if it is in excess.

6. Solution:
   - M_r $Ca(NO_3)_2$ = (1 × 40.1) + (2 × 14.0) + (6 × 16.0) = 164.1.
   - Moles $Ca(NO_3)_2$ = 0.2 ÷ 164.1 = 0.00122 mol (3 significant figures).
   - The chemical equation indicates a ratio of $Ca(NO_3)_2$: Gas of 2 : 5, therefore the theoretical molar yield of gas = 0.00122 × (5 ÷ 2) = 0.00305 mol (3 significant figures).
   - Theoretical molar yield of gas as a volume at r.t.p = 0.00305 × 24 = 0.0731 dm^3 = 73.1 cm^3 (3 significant figures).
   - Percentage Yield = (Actual Yield ÷ Theoretical Yield) × 100 = (48 ÷ 73.1) × 100 = 65.6% (3 significant figures).

7. Solution:
   - M_r $KClO_3$ = (1 × 39.1) + (1 × 35.5) + (3 × 16.0) = 122.6.
   - Moles $KClO_3$ = 1 ÷ 122.6 = 0.00816 mol (3 significant figures).
   - The chemical equation indicates a ratio of $KClO_3$; O_2 of 2 :3, therefore the theoretical molar yield of O_2 = 0.00816 × (3 ÷ 2) = 0.0122 mol (3 significant figures).
   - Theoretical molar yield of O_2 as a volume at r.t.p =0.0122 × 24 = 0.294 dm^3 = 294 cm^3 (3 significant figures).
   - Percentage Yield = (Actual Yield ÷ Theoretical Yield) × 100 = (221 ÷ 294) × 100 = 75.3% (3 significant figures).

8. Solution:
   - M_r MnO_2 = (1 × 54.9) + (2 × 16.0) = 86.9.
   - Moles MnO_2 = 2 ÷ 86.9 = 0.0230 mol (3 significant figures).
   - Moles HCl = 1.1 × 0.025 = 0.0275 mol.
   - Since the ratio of MnO_2 : HCl in the chemical equation is 1 : 4, 0.0230 moles of MnO_2 could react with up to 4 × 0.0230 = 0.0921 moles of HCl (there is **not** enough HCl), whereas 0.0275 moles of HCl can react with 0.0275 ÷ 4 = 0.006875 moles of MnO_2 (there is more than enough MnO_2). Therefore, HCl is the limiting reactant, and MnO_2 is the excess reactant.
   - Since the ratio of HCl : Cl_2 in the chemical equation is 4 : 1, the theoretical yield of Cl_2 is 0.0275 ÷ 4 = 0.006875 moles.
   - Theoretical yield of Cl_2 as a volume at r.t.p = 0.006875 × 24 = 0.165 dm^3 = 165 cm^3.
   - Actual Yield = Theoretical Yield × (38 ÷ 100) = 62.7 cm^3.

9. Solution:
   - M_r $Mg(NO_3)_2$ = (1 × 24.3) + (2 × 14.0) + (6 × 16.0) = 148.3.

- Actual Yield Moles Mg(NO$_3$)$_2$ = 3.9 ÷ 148.3 = 0.0263 mol (3 significant figures).
- Theoretical Yield Moles Mg(NO$_3$)$_2$ = Actual Yield × 100 ÷ 79 = 0.0333 mol (3 significant figures).
- The chemical equation indicates a ratio of Mg(NO$_3$)$_2$: HNO$_3$: of 1 : 2, therefore the theoretical molar amount of HNO$_3$ = 0.0333 × 2 = 0.0666 mol (3 significant figures).
- Concentration = Moles ÷ Volume = 0.0666 ÷ 0.200 = 0.333 mol dm^{-3} (3 significant figures).

10. Solution:
- M$_r$ Mg(NO$_3$)$_2$ = (1 × 24.3) + (2 × 14.0) + (6 × 16.0) = 148.3.
- Actual Yield Moles Mg(NO$_3$)$_2$ = 4.3 ÷ 148.3 = 0.0290 mol (3 significant figures).
- Theoretical Yield Moles Mg(NO$_3$)$_2$ = Actual Yield × 100 ÷ 81 = 0.0358 mol (3 significant figures).
- The chemical equation indicates a ratio of Mg(NO$_3$)$_2$: HNO$_3$: of 1 : 2, therefore the theoretical molar amount of HNO$_3$ = 0.0358 × 2 = 0.0716 mol (3 significant figures).
- Volume = Moles ÷ Concentration = 0.0716 ÷ 1.1 = 0.0651 dm^3 = 65.1 cm^3 (3 significant figures).

Chapter 7: Electrolysis Calculations

In electrolysis an electric current is passed through an **electrolyte**, which is either a molten ionic substance or a solution containing ions. Positive ions (known as **cations**) are attracted to the negative electrode (known as the **cathode**) where they gain electrons and are **reduced**. Negative ions (known as **anions**) are attracted to the positive electrode (known as the **anode**) where they lose electrons and are **oxidized**.

Electrolysis is used for several purposes:
- For the production of many metals (especially reactive metals such as sodium and aluminium)
- For the purification of many metals.
- For the production of some non-metals (especially reactive non-metals such as fluorine and chlorine).
- For coating one metal with another. This process is known as **electroplating**.

Moles of Electrons and the Faraday Constant

Let us consider the electrolysis of molten magnesium chloride ($MgCl_2$). In this case, the electrolyte contains Mg^{2+} and Cl^- ions, and the following reactions (known as **half equations**) will take place at the electrodes (e^- is used to indicate an electron):
- At the cathode: $Mg^{2+}(l) + 2e^- \rightarrow Mg(l)$
- At the anode: $2Cl^-(l) \rightarrow Cl_2(g) + 2e^-$

You will notice that if we were able to calculate the number of moles of electrons flowing through the circuit, we would be able to calculate the amount of magnesium metal (Mg) and chlorine gas (Cl_2) produced as there is a simple ratio between the number of electrons and each of the products produced by electrolysis. But how do we calculate moles of electrons?

The answer is simple:
- The elementary charge of an electron is known to be $1.60217662 \times 10^{-19}$ C.
- There are $6.02214076 \times 10^{23}$ (remember Avogadro's constant from Chapter 1?) electrons in 1 mole.
- Therefore, the charge on 1 mole of electrons = $1.60217662 \times 10^{-19} \times 6.02214076 \times 10^{23}$ = 96485.33289 C mol^{-1} (this is known as the **Faraday constant** (symbol: **F**) or **1 faraday** – for further discussion of this constant/unit see Chapter 8).

Since it is possible to work out the total charge following through a circuit using Charge (Coulombs) = Current (Amps) × Time (seconds), it is therefore possible to calculate how many moles of electrons pass through the circuit using Moles electrons = Charge ÷ 96485.33289, and then by reviewing the ratio between electrons and products in the half equations, calculate how many moles of each of the products are produced.

Example: A current of 5A is passed through an electrolyte of molten magnesium chloride ($MgCl_2$) for a period of 38 minutes, what are the theoretical yields of magnesium (by mass) and chlorine (by gas volume at r.t.p)?

The half equations occurring at the electrodes are:
- At the cathode: $Mg^{2+}(l) + 2e^- \rightarrow Mg(l)$
- At the anode: $2Cl^-(l) \rightarrow Cl_2(g) + 2e^-$

Solution:
- Time = 38 minutes = 38 × 60 = 2280 s.
- Charge = 5 × 2280 = 11400 C.
- Moles of electrons = 11400 ÷ 96485.33289 = 0.118 mol (3 significant) figures.
- As the cathode half equation indicates a ratio of e^- : Mg of 2 : 1, therefore moles Mg = 0.118 ÷ 2 = 0.0591 mol (3 significant figures).
- A_r Mg = 24.3.
- Theoretical yield Mg = 0.0591 × 24.3 = 1.44 g (3 significant figures).
- As the anode half equation indicates a ratio of e^- : Cl_2 of 2 : 1, therefore moles Cl_2 = 0.118 ÷ 2 = 0.0591 mol (3 significant figures).
- Theoretical yield Cl_2 at r.t.p = 0.0591 × 24 = 1.42 dm^3 = 1420 cm^3 (3 significant figures).

Special Considerations for Electroplating
When using electrolysis to perform electroplating, a common scenario is to deposit pure metal at the cathode, but for the anode which is made of an impure metal to gradually dissolve. In such cases, the gain in mass of the cathode will follow a typical electrolysis calculation, but the anode will generally lose slightly more mass, since impurities in the anode will usually fall out and sink to the bottom of the container as the anode dissolves.

Questions
1. At student attempts the electrolysis of tin (II) chloride solution. The half equation that takes place at the cathode is: $Sn^{2+}(aq) + 2e^- \rightarrow Sn(s)$. The half equation that takes place at the anode is $2Cl^-(l) \rightarrow Cl_2(g) + 2e^-$. If the student uses a 3A current for 1 minute, how much tin would be produced, and what volume of chlorine assuming r.t.p?
2. A spoon is electroplated by using it as the cathode while passing current through a copper (II) sulfate solution. The half equation that takes place at the cathode is: $Cu^{2+}(aq) + 2e^- \rightarrow Cu(s)$. If a 3A current is used for 5 minutes, what would be the theoretical yield of copper (by mass) to be deposited on the spoon?
3. Another spoon is electroplated by using it as the cathode while passing current through a silver nitrate sulfate solution. The half equation that takes place at the cathode is: $Ag^+(aq) + e^- \rightarrow Ag(s)$. If

a 3A current is used for 4 minutes, what would be the theoretical yield of silver (by mass) to be deposited on the spoon?

4. A student wishes to electroplate a fork with exactly 1 g of silver. The student does this using silver nitrate solution and electrolysis. The half equation that takes place at the cathode is: $Ag^+(aq) + e^- \rightarrow Ag(s)$. If the student is using a 3A current, how long should he connect the circuit?

5. A student electroplates a knife by using it as the cathode while passing current through a copper (II) sulfate solution. The half equation that takes place at the cathode is: $Cu^{2+}(aq) + 2e^- \rightarrow Cu(s)$. The student connected the circuit for 15 minutes, but unfortunately forgot to record the current. The student however weighed the knife before and after, and determined that the mass of the knife had increased by 0.8 g. What current did the student use?

6. When nitric acid (HNO_3) is electrolyzed, hydrogen (H_2) gas is produced at the cathode and oxygen (O_2) gas at the anode. The half equation at the cathode is: $2H^+(aq) + 2e^- \rightarrow H_2(g)$. The half equation at the anode is $4OH^-(aq) \rightarrow 2H_2O(l) + O_2(g) + 4e^-$. Assuming a current of 2.5 A for 3 minutes, and r.t.p conditions, what volumes of hydrogen and oxygen will be produced?

7. A student electrolyzes nitric acid (HNO_3) producing hydrogen (H_2) gas is produced at the cathode and oxygen (O_2) gas at the anode. The half equation at the cathode is: $2H^+(aq) + 2e^- \rightarrow H_2(g)$. The half equation at the anode is $4OH^-(aq) \rightarrow 2H_2O(l) + O_2(g) + 4e^-$. The student uses a 2.5 A current and produces 1 dm³ of hydrogen at r.t.p. How long did she connect the circuit? What volume of oxygen did she produce in the same period (also at r.t.p)?

8. The electrolysis of molten sodium chloride (NaCl) can be used to produce sodium metal and chlorine gas. The half equation that takes place at the cathode is $Na^+(l) + e^- \rightarrow Na(l)$. The half equation that takes place at the anode is $2Cl^-(l) \rightarrow Cl_2(g) + 2e^-$. A plant produces 1 ton of sodium per day. Assuming the plant is operating continuously, what current was used?

9. Aluminium is produced by electrolysis of molten aluminium oxide (Al_2O_3). The half equation that takes place at the cathode is $Al^{3+}(l) + 3e^- \rightarrow Al(l)$. If a plant produces 2000 tons of aluminium per day, what current is needed?

10. At the same aluminium plant, the half equation at the anode is $O^{2-} \rightarrow [O] + 2e^-$ (where [O] indicates an oxygen atom). Because of the high temperatures the oxygen atoms produced react with the graphite electrode (which must continuously be replaced) according to the following reaction $C + 2[O] \rightarrow CO_2$. What mass of graphite electrodes are used up per day?

Answers to Chapter 7 Questions

1. Solution:
    - Time = 1 × 60 = 60 s.
    - Charge = Current × Time = 3 × 60 = 180 C.
    - Moles of electrons = 180 ÷ 96485.33289 = 0.00187 mol (3 significant figures).
    - The cathode half equation indicates a ratio of e^- : Sn of 2 : 1, therefore moles Sn = 0..00187 ÷ 2 = 0.000933 mol (3 significant figures).
    - A_r Sn = 118.7.
    - Theoretical yield Sn = 0.000933 × 118.7 = 0.111 g (3 significant figures).
    - The anode half equation indicates a ratio of e^- : Cl_2 of 2 : 1, therefore moles Cl_2 = 0..00187 ÷ 2 = 0.000933 mol.
    - Theoretical yield Cl_2 at r.t.p = 0.000933 × 24 = 0.0224 dm^3 = 22.4 cm^3 (3 significant figures).

2. Solution:
    - Time = 5 × 60 = 300 s.
    - Charge = Current × Time = 3 × 300 = 900 C.
    - Moles of electrons = 900 ÷ 96485.33289 = 0.00933 mol (3 significant figures).
    - The cathode half equation indicates a ratio of e^- : Cu of 2 : 1, therefore moles Cu = 0..00933 ÷ 2 = 0.00466 mol (3 significant figures).
    - A_r Cu = 63.5.
    - Theoretical yield Cu = 0.00466 × 63.5 = 0.296 g (3 significant figures).

3. Solution:
    - Time = 4 × 60 = 240 s.
    - Charge = Current × Time = 3 × 240 = 720 C.
    - Moles of electrons = 720 ÷ 96485.33289 = 0.00746 mol (3 significant figures).
    - The cathode half equation indicates a ratio of e^- : Ag of 1 : 1, therefore moles Ag = 0.00746 mol (3 significant figures).
    - A_r Ag = 107.9.
    - Theoretical yield Ag = 0.00746 × 107.9 = 0.805 g (3 significant figures).

4. Solution:
    - A_r Ag = 107.9.
    - Moles of Ag = 1 ÷ 107.9 = 0.00927 mol (3 significant figures).
    - The cathode half equation indicates a ratio of e^- : Ag of 1 : 1, therefore moles of electrons = 0.00927 mol (3 significant figures).
    - Charge required = 0.00927 × 96485.33289 = 894 C (3 significant figures).
    - Time = Charge ÷ Current = 894 ÷ 3 = 298 s = 4 minutes 58 seconds.

5. Solution:
    - A_r Cu = 63.5.
    - Moles Cu = 0.8 ÷ 63.5 = 0.0126 mol (3 significant figures).
    - The cathode half equation indicates a ratio of e^- : Cu of 2 : 1, therefore moles of electrons = 0.0126 × 2 = 0.0252 mol (3 significant figures).

- Charge required = 0.0252 × 96485.33289 = 2430 C (3 significant figures).
- Time = 15 × 60 = 900 s.
- Current = Charge ÷ Time = 2430 ÷ 900 = 2.70 A (3 significant figures).

6. Solution:
   - Time = 3 × 60 = 180 s.
   - Charge = Current × Time = 2.5 × 180 = 450 C.
   - Moles of electrons = 450 ÷ 96485.33289 = 0.00466 mol (3 significant figures).
   - The cathode half equation indicates a ratio of e⁻ : H_2 of 2 : 1, therefore moles H_2 = 0.00466 ÷ 2 = 0.00233 mol (3 significant figures).
   - Theoretical yield H_2 at r.t.p = 0.00233 × 24 = 0.0560 dm³ = 56.0 cm³ (3 significant figures).
   - The anode half equation indicates a ratio of e⁻ : O_2 of 4 : 1, therefore moles O_2 = 0.00466 ÷ 4 = 0.00117 mol (3 significant figures).
   - Theoretical yield O_2 at r.t.p = 0.00117 × 24 = 0.0280 dm³ = 28.0 cm³ (3 significant figures).

7. Solution:
   - Moles H_2 = 1 ÷ 24 = 0.0417 mol (3 significant figures).
   - The cathode half equation indicates a ratio of e⁻ : H_2 of 2 : 1, therefore moles of electrons = 0.0417 × 2 = 0.0833 mol (3 significant figures).
   - Charge = 0.0833 × 96485.33289 = 8040 C (3 significant figures).
   - Time = Charge ÷ Current = 8040 ÷ 2.5 = 3216 seconds = 53 minutes 36 seconds.
   - The anode half equation indicates a ratio of e⁻ : O_2 of 4 : 1, therefore moles of O_2 = 0.0833 ÷ 4 = 0.0208 mol (3 significant figures).
   - Theoretical yield O_2 at r.t.p = 0.0208 × 24 = 0.5 dm³.

8. Solution:
   - Mass Na = 1 ton = 1000000 g.
   - A_r Na = 23.0.
   - Moles Na = 1000000 ÷ 23.0 = 43500 mol (3 significant figures).
   - The cathode half equation indicates a ratio of e⁻ : Na of 1 : 1, therefore moles of electrons = 43500 mol (3 significant figures).
   - Charge = 43500 × 96485.33289 = 4195014473 C.
   - Time = 24 hours = 86400 s.
   - Current = Charge ÷ Time = 4195014473 ÷ 86400 = 48600 A (3 significant figures).

9. Solution:
   - Mass Al = 2000 tons = 2000000000 g.
   - A_r Al = 27.0.
   - Moles Al = 2000000000 ÷ 27.0 = 74100000 mol (3 significant figures).
   - The cathode half equation indicates a ratio of e⁻ : Al of 3 : 1, therefore moles of electrons = 74100000 × 3 = 222000000 mol (3 significant figures).
   - Charge = 222000000 × 96485.33289 = 2.15 × 10¹³ C.
   - Time = 24 hours = 86400 s.
   - Current = Charge ÷ Time = (2.15 × 10¹³) ÷ 86400 = 248000000 A (3 significant figures).

10. Solution:
   - The anode half equation indicates a ratio of e⁻ : [O] of 2 : 1, therefore moles of [O] = 222000000 ÷ 2 = 111000000 mol (3 significant figures).
   - The equation with graphite indicates a ratio of [O] : C of 2 : 1, therefore moles of C = 111000000 ÷ 2 = 55600000 mol (3 significant figures).
   - A_r C = 12.0.
   - Mass of C = 55600000 × 12 = 667000000g = 667 tons (3 significant figures).

Chapter 8: Other Molar Constants & Units

In this chapter, we will review other units that are based on moles.

Faraday Constant and Faradays

As already described in Chapter 7, the charge on 1 mole of electrons is a known as the **Faraday constant** (symbol: **F**) and is named after Michael Faraday (September 22nd, 1791 to August 25th, 1867) who was one of the first scientists to investigate electromagnetism and electrochemistry.

Michael Faraday:

The Faraday constant can be calculated by multiplying the Avogadro constant (the number of particles in a mole) by the elementary charge of an electron. The Faraday constant has a value of approximately 96485.33289 C mol^{-1}.

Although **not** part of the International System of Units (also known as the SI Unit system), it is also possible to use **faradays** as a unit to measure charge. A charge of 1 faraday would be equivalent to 96485.33289 C, a charge of 2 faradays would be equivalent to a charge of 2 × 96485.33289 =192970.6658 C, and so on.

Important Note: Do **not** confuse faradays with **farads** (also symbol: **F**) which are an unrelated unit used for measuring electrical capacitance.

Einsteins

The **einstein** (symbol: **E**) is a unit defined as a mole of photons and is named after the physicist Albert Einstein (March 14th, 1879 to April 18th, 1955). Since einsteins are **not** part of the International System of Units (also known as the SI Unit system), and einsteins redundant with moles, the use of einstein as a unit is probably best avoided.

Nevertheless, you may encounter einsteins in the literature of some scientific fields. For example, photosynthetically active radiation (radiation that plants can use for photosynthesis) is sometimes expressed in microeinsteins per square meter per second ($\mu E\ m^{-2}\ s^{-1}$) – even though that would be exactly equivalent to expressing it in micromoles per square meter per second ($\mu mol\ m^{-2}\ s^{-1}$).

Kilogram-moles and Pound-moles
If you did all the questions in this book, you may have noticed that dealing with quantities in grams and moles can be inconvenient when discussing industrial processes that work with large quantities. For this reason, chemical engineers often preferred to work in kilograms and defined a **kilogram-mole** (symbol: **kg-mol**) as the number of carbon atoms in 12 kg of ^{12}C. 1 kg-mol is exactly equivalent to 1000 "normal" moles (known to chemical engineers as **gram-moles** of **g-mol**), or using modern SI Unit prefixes, 1 kilomole (symbol: kmol). Additionally, concentrations expressed in kg-mol m^{-3} (or kmol m^{-3}) are numerically equivalent to concentrations expressed in mol dm^{-3}, but the former has the advantage of using SI Units which is helpful when calculating flowrates and other matters pertaining to chemical engineering.

Likewise, to avoid the inconvenience of conversions when working with imperial units (or American customary units), some engineers defined a **pound-mole** (symbol: **lb-mol** or **lbmol**) as the number of carbon atoms in 12 lb of ^{12}C. In this case, since 1 pound is to 453.59237 grams, 1 lb-mol = 453.59237 mol.

Summary:
- 1 kg-mol = 1 kmol = 1000 mol.
- 1 lb-mol = 453.59237 mol.

Chapter 9: Mole Factoids

And now some interesting trivia about moles...

No, not that kind of mole!

Mole Day

Mole Day is an unofficial holiday celebrated by chemists, chemistry students, and their supporters around the world.

- Mole Day falls on October 23rd of each year and is celebrated from 6:02am to 6:02pm. This corresponds to 6:02 10/23 (if writing the date in the American style), and Avogadro's number, you will recall, is approximately 6.02×10^{23}.
- The American Chemical Society (ACS) – official website http://www.acs.org/ – sponsors National Chemistry Week in the week of October during which October 23rd falls.

10:23 Campaign

The 10:23 Campaign – official website http://www.1023.org.uk/ – is a campaign against homoeopathy (a system of alternative medicine), originally organized by the Merseyside Skeptics Society in the United Kingdom, but subsequently supported up by numerous skeptic groups.

Homoeopathic remedies are based on chemicals diluted to such an extreme degree that **not** even a single molecule of the supposedly active ingredient is likely to remain in the patient's dose of the remedy. On several occasions, skeptics have drawn attention to this fact, by "overdosing" (taking many times the recommended dose) of various homoeopathic remedies. The 10:23 campaign is named after the 10^{23} in Avogadro's number.

Conclusion

Well done for finishing the book! I hope you enjoyed it.

For more moles and stoichiometry resources, please go to:

- http://www.suniltanna.com/moles

To find out about other educational books that I have written, please go to:

- For chemistry books: http://www.suniltanna.com/chemistry
- For science books: http://www.suniltanna.com/science
- For math books: http://www.suniltanna.com/math

Remember: If you enjoyed this book or it helped you, please post a positive review on Amazon!

Reference – Important Constants & Formulae

Avogadro Constant
$6.02214076 \times 10^{23}$

Volumes
$1 \text{ dm}^3 = 1{,}000 \text{ cm}^3$
$1 \text{ m}^3 = 1{,}000 \text{ dm}^3 = 1{,}000{,}000 \text{ cm}^3$

Moles & Mass:
Moles of atoms = Mass (grams) ÷ A_r
Moles of molecules = Mass (grams) ÷ M_r
Moles of formula units = Mass (grams) ÷ M_r

Moles & Solutions
Moles = Concentration (mol dm^{-3}) × Volume (dm^3)
Moles = Concentration (g dm^{-3}) × Volume (dm^3) ÷ M_r
Mass (g) = Concentration (mol dm^{-3}) × Volume (dm^3) × M_r
Mass (g) = Concentration (g dm^{-3}) × Volume (dm^3)
Concentration (mol dm^{-3}) = Concentration (g dm^{-3}) ÷ M_r
Titration: Concentration1 (mol dm^{-3}) × Volume1 (dm^3) ÷ n1 = Concentration2 (mol dm^{-3}) × Volume2 (dm^3) ÷ n2

Moles & Ideal Gases
Room Temperature and Pressure (r.t.p) = 25°C (298.15K), 101.325 kPa
At r.t.p: Moles = Volume (dm^3) ÷ 24
Standard Temperature and Pressure (s.t.p) = 0°C (273.15K), 101.325 kPa
At s.t.p: Moles = Volume (dm^3) ÷ 22.4
Ideal Gas Law: Moles = (Pressure (Pa) × Volume (m^3)) ÷ (R (J K^{-1} mol^{-1}) × Temperature (K))
Where Ideal Gas Constant = R = 8.314 J K^{-1} mol^{-1}

Moles & Electrolysis
Moles of Electrons = Charge (Coulombs) ÷ 96485.33289
Where 96485.33289 C mol^{-1} = Faraday constant = Charge of 1 mole of electrons
Charge (C) = Current (Amps) × Time (seconds)

Percentage by Mass
% By Mass = (Total Relative Mass of Atoms of an Element ÷ M_r of Compound) × 100

Yield
% Yield = (Actual Yield ÷ Theoretical Yield) × 100

Atom Economy
Atom Economy = (Relative Mass of Desired Products ÷ Relative Mass of All Products) × 100
or: Atom Economy = (Relative Mass of Desired Products ÷ Relative Mass of All Reactants) × 100

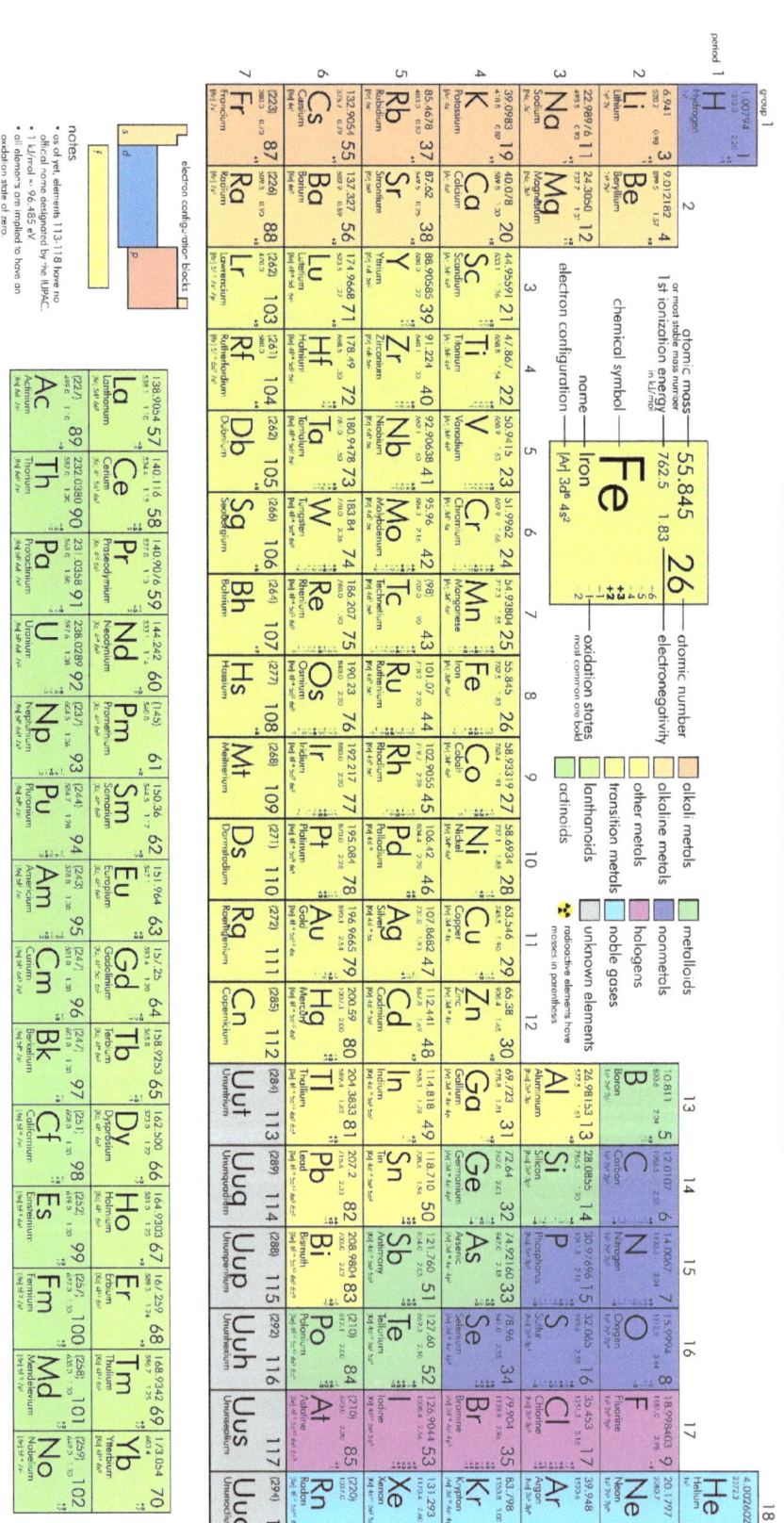

www.ingramcontent.com/pod-product-compliance
Lightning Source LLC
Chambersburg PA
CBHW051158220526
45473CB00003B/814